COMPLEX ANALYSIS

AN INVITATION

T0220412

A Concise Introduction to Complex Function Theory

COMPLEX ANALYSIS
AN INVITATION

Murali Rao
Department of Mathematics
University of Florida
USA

Henrik Stetkær
Institute of Mathematics
University of Aarhus
Denmark

World Scientific
Singapore • New Jersey • London • Hong Kong

Published by

World Scientific Publishing Co. Pte. Ltd.
P O Box 128, Farrer Road, Singapore 9128
USA office: 687 Hartwell Street, Teaneck, NJ 07666
UK office: 73 Lynton Mead, Totteridge, London N20 8DH

Library of Congress Cataloging-in-Publication data is available.

COMPLEX ANALYSIS. An Invitation

ISBN 981-02-0375-6
      981-02-0376-4 (pbk)

Printed in Singapore by JBW Printers & Binders Pte. Ltd.

Preface

This textbook is a rigorous introduction to the theory of functions of one complex variable. Much of the material has been used in both graduate and undergraduate courses at Aarhus University. It is our impression that courses in complex function theory these days at many places are postponed to leave space for other topics like point set topology and measure theory. Thus we have felt free to assume that the students as background have some point set topology and calculus of several variables, and that they understand $\epsilon\delta$ arguments. From Chapter X on we even use Lebesgue's dominated and monotone convergence theorems freely. But the proofs are meant for the students and so they are fairly detailed.

We have made an effort to whenever possible to give references to literature that is accessible for the students and/or puts the theory in perspective, both mathematically and otherwise. E.g. to the Chauvenet prize winning paper [Za] and to the fascinating gossip on Bloch's life in [Ca] and its sequel [CF]. Our goal is not to compete with the existing excellent textbooks for historical notes and remarks or for wealth of material. For that we refer the interested reader to, say the monumental work [Bu] and the classic [SG].

We have tried to reach some of the deeper and more interesting results (Picard's theorems, Riemann's mapping theorem, Runge's approximation theorems) rather early, and nevertheless to give the very basic theory an adequate treatment.

Standard notation is enforced throughout. A possible exception is that B[a,r] is the closed ball with center a and radius r in analogy with the notation for a closed interval.

An important part of any course is the set of exercises. We have exercises after each chapter. They are meant to be doable for the students, so we have quite often provided hints about how to proceed.

A couple of times we have succumbed to the temptation of making a digression to an interesting topic, that will not be pursued, e.g. Tauber's theorem, Hadamard's gap theorem and the prime number theorem. We hope that the reader will be irresistibly tempted as well.

The authors welcome correspondence with criticism and suggestions. In particular about literature on the level of students that are about to start their graduate studies in mathematics.

Contents

Contents

Chapter 1 Power Series

Section 1 Elementary facts

This section contains basic results about convergence of power series. The simplest and most important example is the *geometric series*

$$\sum_{k=0}^{\infty} z^k = \frac{1}{1-z}$$

which converges absolutely for any z in the open unit disc. It emerges in so many other contexts than complex function theory that it may be considered one of the fundamental elements of mathematics (See the thought-provoking paper [Ha]).

Definition 1.

A power series around $z_0 \in \mathbf{C}$ *is a formal series of the form*

$$(1) \quad \sum_{k=0}^{\infty} a_k (z - z_0)^k$$

where the coefficients $a_k \in \mathbf{C}$ *are fixed and where* $z \in \mathbf{C}$.

We shall very often only consider the case $z_0 = 0$, i.e.

$$(2) \quad \sum_{k=0}^{\infty} a_k z^k$$

because it will be obvious how to derive the general case from this more handy special case.

We shall in Chapter VI,§1 encounter Laurent series, i.e. "power" series in which the summation ranges over $\mathbf{Z}$, not just $\mathbf{N}$. A Laurent series will be treated as a sum of a power series in z and another in the variable $1/z$.

Proposition 2.

Consider the power series (2), and let us assume that the set $\left\{ a_k \zeta^k \mid k = 0, 1, 2, \cdots \right\}$ *is bounded for some* $\zeta \in \mathbf{C}$.

Then for any $\rho < |\zeta|$ *the power series (2) converges absolutely and uniformly in the closed disc* $B[0,1] = \{ z \in \mathbf{C} \mid |z| \le \rho \}$.

Proof: Because of boundedness $|a_k| |\zeta^k| \le M$ for some M and all k. Thus if $|z| \le \rho < |\zeta|$, then

$$\left| a_k z^k \right| \le |a_k| |\zeta|^k \rho^k |\zeta|^{-k} \le M \left(\frac{\rho}{|\zeta|} \right)^k$$

1

So the series is dominated by a convergent series (the geometric) and the proposition follows from Weierstrass' M-test. $\qquad\square$

Proposition 2 says that the power series (2) either converges for all complex z or there is a number m such that $\sup_k |a_k m^k| = \infty$. We define the *radius of convergence* $\rho \in [0, \infty]$, of the power series (1) by

$$(3) \quad \rho := \inf \left\{ m > 0 \mid \sup_k \left| a_k m^k \right| = \infty \right\}$$

and the *circle of convergence* as $\{z \in \mathbf{C} \mid |z - z_0| = \rho\}$, provided $\rho < \infty$. The reason for this terminology is that (1) converges at each interior point of the circle of convergence, and diverges at each exterior point.

A side remark: If z is not a complex number but, say a matrix, then the series may converge even if the norm of z is bigger than ρ; that can happen in exponentiating a matrix.

Theorem 3.

Let ρ denote the radius of convergence of the power series (1).

(a) (The Cauchy-Hadamard radius of convergence formula)

$$(4) \quad \rho = \frac{1}{\displaystyle\limsup_{k \to \infty} |a_k|^{\frac{1}{k}}}$$

(b) (The quotient formula). If $a_n \neq 0$ for all n, then

$$\rho = \lim_{n \to \infty} \left| \frac{a_n}{a_{n+1}} \right|$$

provided that the limit exists in $[0, \infty]$.

(c) (1) converges for any $R < \rho$ uniformly on the disc $B[z_0, R]$, and it converges absolutely on $B(z_0, \rho)$.

Proof: (a) We prove that

$$\limsup_{k \to \infty} |a_k|^{\frac{1}{k}} \geq \frac{1}{\rho}$$

by contradiction: If

$$\limsup_{k \to \infty} |a_k|^{\frac{1}{k}} < \frac{1}{\rho}$$

then there exists $m > \rho$ such that

$$\limsup_{k \to \infty} |a_k|^{\frac{1}{k}} < \frac{1}{m}$$

2

So except for finitely many k we have $m|a_k|^{\frac{1}{k}} < 1$, i.e. $\left|a_k m^k\right|$ is bounded, contradicting the definition of ρ.

Next we shall prove that

$$\limsup_{k\to\infty} |a_k|^{\frac{1}{k}} \le \frac{1}{\rho}$$

or equivalently

$$\limsup_{k\to\infty} |a_k|^{\frac{1}{k}} \le \frac{1}{b}$$

for any $b < \rho$. In this case we note that there exists an $M > 0$ such that $|a_k| b^k \le M$ for all k. Hence

$$\limsup_{k\to\infty} |a_k|^{\frac{1}{k}} b \le \lim_{k\to\infty} M^{\frac{1}{k}} = 1$$

(b) is left to the reader as Exercise 2.

(c) is part of Proposition 2. $\square$

As examples we note that the geometric series has radius of convergence 1, and that the series

$$\sum_{n=0}^{\infty} \frac{z^n}{n!}$$

has radius of convergence $\rho = \infty$.

Corollary 4.

If the power series $\sum_{k=0}^{\infty} a_k z^k$ *has radius of convergence* ρ, *then so does the formally differentiated series* $\sum_{k=0}^{\infty} a_{k+1}(k+1)z^k$.

Proof: Note that $\sum a_{k+1}(k+1)z^k$ converges iff $\sum a_k k z^k$ does. Then use Cauchy-Hadamard and that $\lim_{k\to\infty} k^{\frac{1}{k}} = 1$. $\square$

Recalling that a uniformly convergent series of continuous functions has continuous sum, we get

Proposition 5.

The sum of a power series is continuous inside its circle of convergence.

Much more is true inside, as we will discover in the next chapter, in which we discuss differentiability. The behavior of the power series at the convergence circle is more delicate, and aspects of it will be treated in §2.

An Nth order polynomial which is 0 at $N+1$ points is identically 0. A primitive extension of this to power series, viewed as polynomials of infinite order, is the following:

Proposition 6.

Assume that the power series (1) has positive radius of convergence, and let f denote its sum.

(α) If $f(z) = 0$ for z in a set which has z_0 as an accumulation point, then the coefficients a_k are all 0.

(β) In particular, f determines the coefficients uniquely.

Our final version of Proposition 6 will be the "Unique Continuation Theorem" (Theorem IV.11).

Proof of the proposition: (β) is an immediate consequence of (α), so we will only do (α). We also take $z_0 = 0$.

Let $z_1, z_2, \cdots$ be a sequence from $\mathbf{C}$, such that $z_k \to 0$ as $k \to \infty$, $z_k \neq 0$ and $f(z_k) = 0$ for $k = 1, 2, \cdots$. By the continuity of f we get

$$a_0 = f(0) = \lim_{k \to \infty} f(z_k) = 0$$

But then

$$\frac{f(z)}{z} = \sum_{k=0}^{\infty} a_{k+1} z^k$$

so replacing $f(z)$ by $f(z)/z$ and by repeating the argument we find that $a_1 = 0$. Proceeding in this way we get successively that $a_n = 0$ for $n = 0, 1, 2, \cdots$. $\square$

For more detailed information on power series we refer the reader to [Kn].

Section 2 The theorems of Abel and Tauber

In this paragraph we will in special cases study the behavior of the power series (2) on its circle of convergence. Unfortunately, in general almost anything can happen. One can here mention a famous example, due to L. Fejér, of a power series that converges uniformly, but not absolutely, on the closed unit disc. (See p. 125 ff of [Za] or p.122 of [HI]). Of course, absolute convergence on the circle of convergence holds either everywhere or nowhere. In case of absolute convergence the power series even converges uniformly on the closed disc $B[0, \rho] = \{z \in \mathbf{C} | |z| \le \rho\}$, where ρ is the radius of convergence. We are thus left with the case of nonabsolute convergence which is technically unpleasant.

Theorem 7 (Abel's theorem).

Assume that the power series

$$f(z) = \sum_{k=0}^{\infty} a_k z^k$$

converges for $z = 1$ and hence for each $|z| \leq 1$.

Then $\sum_{k=0}^{\infty} a_k\, x^k$ converges uniformly to $f(x)$ on the closed interval [0,1]. In particular, f is continuous on [0,1] , so

$$f(x) \to \sum_{k=0}^{\infty} a_k \ as \ x \to 1 \ in \ [0,1]$$

Proof: Note that the sequence

$$r_k := - \sum_{m>k}^{\infty} a_m \ , \ k = 0, 1, 2, \cdots$$

converges to 0 as $k \to \infty$. In particular it is bounded so that the series $\sum_{k=0}^{\infty} r_k z^k$ converges whenever $|z| \leq 1$. Now, for any z such that $|z| \leq 1$ and any natural number N we have

$$f(z) - \sum_{k=0}^{N} a_k z^k = \sum_{k=N+1}^{\infty} a_k z^k = \sum_{k=N+1}^{\infty} (r_k - r_{k-1}) z^k$$

$$= \sum_{k=N+1}^{\infty} r_k z^k - \sum_{k=N+1}^{\infty} r_{k-1} z^k$$

$$= \sum_{k=N+1}^{\infty} r_k z^k (1 - z) - r_N z^{N+1}$$

from which we for $z = x \in [0,1]$ get the estimate

$$\left| f(x) - \sum_{k=0}^{N} a_k x^k \right| \leq \sup_{k>N} \{|r_k|\} \sum_{k=N+1}^{\infty} x^k (1-x) + |r_N| \leq \sup_{k>N} \{|r_k|\} + |r_N|$$

Thus we have for any $x \in [0,1]$ that

$$\left| f(x) - \sum_{k=0}^{N} a_k z^k \right| \leq 2 \sup_{k \geq N} \{|r_k|\}$$

the case $x = 1$ being trivial. □

Examples 8.

A combination of Abel's theorem with the identities

$$\text{Log}(1+z) = \sum_{k=1}^{\infty} \frac{(-1)^{k-1}}{k} z^k \quad \text{for } |z| < 1$$

$$\arctan z = \sum_{k=0}^{\infty} \frac{(-1)^k}{2k+1} z^{2k+1} \quad \text{for } |z| < 1$$

yields the following pretty formulas

$$\text{Log } 2 = 1 - \frac{1}{2} + \frac{1}{3} - \frac{1}{4} + \cdots$$

$$\frac{\pi}{4} = 1 - \frac{1}{3} + \frac{1}{5} - \frac{1}{7} + \cdots$$

which are called respectively *Brouncker's series* and *Leibniz' series*.

Our next theorem provides a partial inverse to Abel's theorem.

Theorem 9 (Tauber's theorem).

If $f(z) = \sum_{k=0}^{\infty} a_k z^k$ *converges on the unit disc,* $ka_k \to 0$ *as* $k \to \infty$ *and* $\lim f(x)$ *exists for* $x \to 1_-$ *, then* $\sum_{k=0}^{\infty} a_k$ *converges.*

Proof: Introducing

$$s_n := \sum_{k=0}^{n} a_k \quad \text{and} \quad \omega(n) := \sup_{k \geq n} \{k|a_k|\}$$

we find since

$$1 - x^k \leq k(1-x) \quad \text{and} \quad \sum_{k=n+1}^{\infty} \frac{x^k}{k} \leq \frac{1}{n+1} \sum_{k=n+1}^{\infty} x^k \leq \frac{1}{(1-x)(n+1)}$$

that

$$(5) \quad |f(x) - s_n| \leq \sum_{k=1}^{n} |a_k| \left(1 - x^k\right) + \omega(n+1) \sum_{k=n+1}^{\infty} \frac{x^k}{k}$$

$$\leq \omega(0)(n+1)(1-x) + \frac{\omega(n+1)}{(1-x)(n+1)}$$

The expression on the right hand side of (5) is of the form $At + Bt^{-1}$ and when $t^2 = B/A$ it equals $2\sqrt{AB}$. So when x_n satisfies

$$(1 - x_n)^2 = \frac{\omega(n+1)}{(n+1)^2 \omega(0)}$$

6

we have from (5) that

$$|f(x_n) - s_n| \le 2\sqrt{\omega(0)\omega(n+1)}$$

which tends to 0 as $n \to \infty$ because $\omega(n) \to 0$. $\square$

Remark.   More difficult is Littlewood's Tauberian theorem which requires only that $\{ka_k\}$ is bounded. For an accessible proof (due to Karamata and Wieland) see p.129 ff of [Za].

Even though the following identity is easy to state and prove it is often useful in manipulations with explicitly given series. We used it, without actually saying so, in the proof of Abel's theorem.

Proposition 10 (Abel's partial summation formula).

Let c_k and d_k , $k = 0, 1, \cdots, n$ be complex numbers , and put $s_k := \sum_{j=0}^{k} d_j$ for $k = 0, 1, \cdots, n$.
Then

$$(6) \quad \sum_{k=0}^{n} c_k d_k = \sum_{k=0}^{n-1} (c_k - c_{k+1})s_k + c_n s_n$$

Section 3 Liouville's theorem

Liouville's theorem (that a bounded entire function is constant), is not only interesting in itself, but it is also useful in many unexpected situations. It has a generalization which says that a power series that grows at most polynomially at ∞ is a polynomial. We present the generalization here:

Theorem 11 (Liouville's theorem).
Let the power series

$$f(z) = \sum_{k=0}^{\infty} a_k z^k$$

have infinite radius of convergence.
(i) If f is bounded then f is constant.
(ii) More generally, if f for some nonnegative constants a, b and m satisfies the estimate

$$|f(z)| \le a|z|^m \ \text{for all} \ |z| > b$$

then f is a polynomial of degree $\le m$.

7

Of course Liouville's theorem does not hold for smooth functions of one or more real variables. For example, the function $f(x) = \sin x$ is a bounded, smooth function of one real variable, but it is certainly not constant.

Liouville's theorem can be generalized to harmonic functions (See Chapter XII. Exercise 10).

Proof: It suffices to prove (ii). The coefficient a_n may for any $n = 0, 1, 2, \cdots$ be found by the formula

$$a_n = \frac{1}{2\pi R^n} \int_0^{2\pi} e^{-in\theta} f\left(Re^{i\theta}\right) d\theta \ \text{ for any } \ R > 0$$

which crops up when we introduce the power series expansion of f on the right hand side. In particular we get the estimate

$$|a_n| \leq \frac{1}{2\pi R^n} \int_0^{2\pi} \left|f\left(Re^{i\theta}\right)\right| d\theta$$

Substituting the estimate for f from (ii) into this we find that

$$|a_n| \leq aR^{m-n} \ \text{ for all } \ R > b$$

Letting $R \to \infty$ we see that $a_n = 0$ whenever $n > m$, so the power series reduces to the polynomial $f(z) = \sum_{n=0}^{N} a_n z^n$, where N is the integer part of m. $\square$

For proof without calculus (!) see [Le].

Section 4 Important power series

In this paragraph we collect the most important power series expansions. Some of them will be derived later. In particular we will discuss the exponential function in the next chapter.

$$\frac{1}{1-z} = \sum_{n=0}^{\infty} z^n \quad \text{for } |z| < 1$$

$$e^z = \sum_{n=0}^{\infty} \frac{z^n}{n!} \quad \text{for } z \in \mathbb{C}$$

$$\cos z = \frac{e^{iz} + e^{-iz}}{2} = \sum_{n=0}^{\infty} \frac{(-1)^n}{(2n)!} \quad \text{for } z \in \mathbb{C}$$

$$\sin z = \frac{e^{iz} - e^{-iz}}{2i} = \sum_{n=0}^{\infty} \frac{(-1)^n}{(2n+1)!} z^{2n+1} \quad \text{for } z \in \mathbb{C}$$

$$\cosh z = \frac{e^z + e^{-z}}{2} = \sum_{n=0}^{\infty} \frac{1}{(2n)!} z^{2n} \quad \text{for } z \in \mathbb{C}$$

$$\sinh z = \frac{e^z - e^{-z}}{2} = \sum_{n=0}^{\infty} \frac{1}{(2n+1)!} z^{2n+1} \quad \text{for } z \in \mathbb{C}$$

$$\text{Log}(1+z) = \sum_{n=1}^{\infty} \frac{(-1)^{n-1}}{n} z^n \quad \text{for } |z| < 1$$

$$(1+z)^\alpha = \sum_{n=0}^{\infty} \binom{\alpha}{n} z^n \quad \text{for } |z| < 1 \quad \text{(principal branch)}$$

$$\arctan z = \sum_{n=0}^{\infty} \frac{(-1)^n}{2n+1} z^{2n+1} \quad \text{for } |z| < 1$$

$$\arcsin z = \sum_{n=0}^{\infty} \frac{1 \cdot 3 \cdot 5 \cdots (2n-1)}{2 \cdot 4 \cdot 6 \cdots 2n} z^{2n+1} \quad \text{for } |z| < 1$$

$$J_\nu(z) = \frac{1}{2\pi} \int_0^{2\pi} e^{i(z\sin t - \nu t)} dt$$

$$= \sum_{n=0}^{\infty} \frac{(-1)^n \left(\frac{z}{2}\right)^{\nu+2n}}{n! \Gamma(\nu+n+1)} \quad \text{for } z \in \mathbb{C} \quad \text{(Bessel function)}$$

Section 5 Exercises

1. Determine the radius of convergence and study the behavior on the circle of convergence of the following four power series :

$$\sum_{k=0}^{\infty} z^k \ , \ \sum_{k=1}^{\infty} \frac{z^k}{k^2} \ , \ \sum_{k=0}^{\infty} \frac{z^k}{k+1} \ , \ \sum_{k=1}^{\infty} \frac{z^{kn}}{k}$$

where n is a positive integer.

2. Let $a_n > 0$ and $b_n \in \mathbb{C}\setminus\{0\}$ for $n = 0, 1, \cdots$.

9

(α) Show that

$$\liminf_{n\to\infty} \frac{a_{n+1}}{a_n} \leq \liminf_{n\to\infty} a_n^{\frac{1}{n}} \leq \limsup_{n\to\infty} a_n^{\frac{1}{n}} \leq \limsup_{n\to\infty} \frac{a_{n+1}}{a_n}$$

(β) Show that the radius of convergence of $\sum\limits_{n=0}^{\infty} b_n z^n$ is $\lim\limits_{n\to\infty} |b_n/b_{n+1}|$, provided that the limit exists in $[0,\infty]$.

3. Let $a \neq 0$. Show that the radius of convergence of the power series

$$1 + az + a(a-2b)\frac{z^2}{2!} + a(a-3b)^2\frac{z^3}{3!} + \cdots$$

is $1/(e|b|)$.

4. Find an example of a power series $\sum\limits_{n=0}^{\infty} a_n z^n$ that converges only at $z=0$.

5. (α) Let $\sum\limits_{n=0}^{\infty} a_n z^n$ and $\sum\limits_{n=0}^{\infty} b_n z^n$ be power series with radii of convergence ρ and σ respectively. Consider the product series $\sum\limits_{n=0}^{\infty} c_n z^n$, defined by

$$c_n := \sum_{k=0}^{n} a_k b_{n-k} \text{ for } n = 0,1,\cdots$$

Show that its radius of convergence τ satisfies $\tau \geq \min(\rho,\sigma)$, and that

$$\sum_{n=0}^{\infty} c_n z^n = \left(\sum_{n=0}^{\infty} a_n z^n\right)\left(\sum_{n=0}^{\infty} b_n z^n\right) \text{ whenever } |z| < \min(\rho,\sigma)$$

(β) Let $\sum\limits_{n=0}^{\infty} a_n$ and $\sum\limits_{n=0}^{\infty} b_n$ be convergent series of complex numbers, and let $\sum\limits_{n=0}^{\infty} c_n$ denote the series given by

$$c_n = \sum_{k=0}^{n} a_k b_{n-k} \text{ for } n = 0,1,\cdots$$

Show that

$$\sum_{n=0}^{\infty} c_n = \left(\sum_{n=0}^{\infty} a_n\right)\left(\sum_{n=0}^{\infty} b_n\right)$$

provided that $\sum\limits_{n=0}^{\infty} c_n$ is convergent.

6. (Eneström-Kakeya)

(A) If $a_n \geq a_{n-1} \geq \cdots \geq a_1 \geq a_0 > 0$, then

$$\sum_{k=0}^{n} a_k z^k \neq 0 \text{ for all } |z| > 1$$

(Hint: Abel's partial summation formula).

(B) Deduce from (A), that $a_n > a_{n-1} > \cdots > a_1 > a_0 > 0$ implies that $\sum\limits_{k=0}^{n} a_k z^k$ has all its roots in the open disc $|z| < 1$.

For extensions of the Eneström-Kakeya result consult e.g. [DG]. The result has applications in digital signal analysis, where it is used in examinations of stability of filters.

7. In this problem we generalize the Alternating Series Theorem from the real domain to the complex domain (point (α)). We consider a power series $\sum\limits_{k=0}^{\infty} a_k z^k$ where $\{a_k\}$ is a sequence of positive numbers decreasing to 0. We let $f(z)$ denote its sum.

(α) Prove Abel's test (due to E. Picard): The series converges everywhere on the unit circle except possibly at $z = 1$.

Hint : Abel's partial summation formula. Or use the geometrical proof in the short note [Be].

(β) Give an example in which the series diverges at $z = 1$ and another in which it converges at $z = 1$.

(γ) Show that $(1 - z)f(z) \to 0$ as $z \to 1_-$.

8. Let $f(z) := \sum\limits_{k=0}^{\infty} a_k z^k$ have radius of convergence $\rho > 0$. Show for each ξ such that $|\xi| = 1$ and such that ξ is not a root of unity [i.e. $\xi^n \neq 1$ for all $n \geq 1$] that

$$\lim_{n \to \infty} \frac{1}{n+1} \sum_{k=0}^{n} f\left(\xi^k z\right) = 0$$

uniformly in z on compact subsets of $B(0, \rho)$.

Deduce the *Cauchy inequality*

$$|a_0| \leq \sup_{|z|=r} |f(z)| \text{ for each } r \in]0, \rho[$$

9. (A generalization of Abel's theorem)

Let us assume that the power series $f(z) := \sum\limits_{k=0}^{\infty} a_k z^k$ converges for $z = 1$ and hence for each $|z| < 1$.

(α) Let $C > 0$. Show that $\sum\limits_{k=0}^{\infty} a_k z^k$ converges to $f(z)$ uniformly on the set

$$\left\{ z \in B(0, 1) \mid \frac{|1 - z|}{1 - |z|} \leq C \right\} \cup \{1\}$$

(β) Let $\theta \in [0, \frac{\pi}{2}[$. Show that $\sum\limits_{k=0}^{\infty} a_k z^k$ converges to $f(z)$ uniformly on the set (sketch it !)

$$\{z \in B(0, 1) | |1 - z| \leq \cos\theta \ , \ |\text{Arg}(1 - z)| \leq \theta\} \cup \{1\}$$

10. Show by dint of an example that the assumption $k a_k \underset{k \to \infty}{\to} 0$ in Tauber's theorem cannot be deleted.

Chapter 2 Holomorphic and Analytic Functions

Section 1 Basics of complex calculus

Definition 1.

Let f be a complex valued function defined on an open set Ω in the complex plane. f is said to be complex differentiable *at $z_0 \in \mathbf{C}$ if the limit*

$$(1) \quad \lim_{h \to 0} \frac{f(z_0 + h) - f(z_0)}{h}$$

exists in $\mathbf{C}$.

The limit is called the complex derivative *of f at z_0 and is denoted $f'(z_0)$ or $\frac{df}{dz}(z_0)$. We say that f is* holomorphic *on Ω if f is complex differentiable at each point of Ω. The function $f' : \Omega \to \mathbf{C}$ is then called the* complex derivative *of f or just the* derivative. *If there exists a holomorphic function F defined on Ω such that $F' = f$, we say that F is a* primitive *of f. If f is holomorphic in all of $\mathbf{C}$ then f is said to be* entire.

Like in real variable theory we find that f is continuous on an open set Ω if it is holomorphic on Ω.

By routine calculations the usual rules for differentiation also hold in the complex case:

Theorem 2.

(a) If f and g are holomorphic on Ω then so are $f + g$ and fg, and $(f + g)' = f' + g'$, $(fg)' = f'g + fg'$.

(b) Let f and g be holomorphic on Ω. If g is not identically 0 on Ω then f/g is holomorphic on the open set $\{z \in \Omega | g(z) \neq 0\}$, and on this set

$$\left(\frac{f}{g}\right)' = \frac{f'g - fg'}{g^2}$$

(c) The chain rule: If f is holomorphic on Ω, g on Ω_1 and $f(\Omega)$ is contained in Ω_1, then the composition $g \circ f$ is holomorphic on Ω, and

$$(g \circ f)'(z) = g'(f(z))f'(z) \quad \text{for all} \ z \in \Omega$$

As is easy to see from the theorem, the set of functions which are holomorphic on Ω forms an algebra over the complex numbers. We denote this algebra $\text{Hol}(\Omega)$.

Examples.

(a) Constants are holomorphic and their derivatives are 0.

13

(b) The function $z \rightarrow z$ is holomorphic and its derivative is 1.

(c) The function $z \rightarrow z^n$ is holomorphic on $\mathbf{C}$ for $n = 0, 1, 2, \cdots$, and on $\mathbf{C} \backslash \{0\}$ for $n = -1, -2, \cdots$. In both cases with derivative $(z^n)' = nz^{n-1}$.

(d) A polynomial $p(z) = \sum\limits_{k=0}^{n} a_k z^k$ is entire and $p'(z) = \sum\limits_{k=1}^{n} k a_k z^{k-1}$.

Theorem 3.

Let $f : \Omega \rightarrow \mathbf{C}$, where Ω is an open subset of the complex plane, and write $f = u + iv$, where u and v are real valued functions.

(a) If f is holomorphic on Ω then all the first order partial derivatives of u and v exist in Ω and the Cauchy-Riemann equations are satisfied there, i.e.

$$\frac{\partial u}{\partial x} = \frac{\partial v}{\partial y} \ \text{ and } \ \frac{\partial u}{\partial y} = -\frac{\partial v}{\partial x}$$

or in a more compact notation

$$\frac{\partial f}{\partial \bar{z}} = 0 \ , \ \text{ where } \ \frac{\partial}{\partial \bar{z}} := \frac{1}{2}\left(\frac{\partial}{\partial x} + i \frac{\partial}{\partial y} \right)$$

(b) If $f \in C^1(\Omega)$ satisfies the Cauchy-Riemann equations in Ω then f is holomorphic in Ω.

Proof:

(a) This is seen from (1) by letting h tend to 0 along the real and the imaginary axes respectively. (b) Left to the reader as Exercise 4. $\qquad\qquad\square$

Remarks on the Cauchy-Riemann equations: The differentiability conditions in point (b) of the theorem can be relaxed considerably. One generalization is the *Looman-Menchoff theorem*:

Let $f \in C(\Omega)$. Then $f \in Hol(\Omega)$, if $\partial f/\partial x$ and $\partial f/\partial y$ exist in all of Ω and satisfy the Cauchy-Riemann equations there (See Theorem 1.6.1 p.48 of [Na]).

Another generalization: It suffices that the Cauchy-Riemann equations are satisfied in the sense of distributions (See [Tr;Theorem 5.1 p.36]). For more information see [GM].

The Cauchy-Riemann equations say roughly speaking that f does not depend on $\bar{z}$, and so it is a function of z only.

If f is holomorphic on an open set Ω, then (use the Cauchy-Riemann equations) $|f'|^2 = |grad\, u|^2 = |grad\, v|^2$, and $|grad\, u|^2 = (\partial u/\partial x)^2 + (\partial u/\partial y)^2$. In particular, $f' = 0$ throughout Ω implies that f is constant on each connected component of Ω.

Not all functions are holomorphic: E.g. $z \rightarrow \bar{z}$ is not.

As we saw in an example above, the functions $f_0 = 1$ and $f_1(z) = z$ are entire. Hence so is each polynomial in z. And each rational function, i.e. each function of the form $f = P/Q$ where P and Q are polynomials, is holomorphic off the zeros of the denominator Q. It turns out (Propositions III.5 and III.20 of the next chapter) that continuous roots and logarithms are holomorphic, too. So are power series; we give a direct proof here. A shorter, but more sophisticated proof is given below following Lemma 12.

Proposition 4.

The sum of a power series is holomorphic inside its circle of convergence, and its derivative can be got by term-by-term differentiation: If

$$f(z) = \sum_{n=0}^{\infty} a_n(z - z_0)^n$$

has radius of convergence $\rho > 0$, then f is holomorphic on $B(z_0, \rho)$ and

$$f'(z) = \sum_{n=1}^{\infty} a_n n(z - z_0)^{n-1} \ for \ z \in B(z_0, \rho)$$

In particular all the complex derivatives $f', f'', \cdots$ exist in $B(z_0, \rho)$.

Proof: Observe that the formula for f' makes sense by Corollary I.4. Let us for convenience of writing assume that $z_0 = 0$.

The technical key to the proof is the following inequality:

$$\left| (z + h)^n - z^n - hnz^{n-1} \right| \leq \left| \frac{h}{\delta} \right|^2 (|z| + \delta)^n$$

valid for $n \in \mathbb{N}$, $h, z \in \mathbb{C}$ such that $0 < |h| \leq \delta$.

Indeed, using the inequality we find with $\delta < \rho - |z|$ that

$$\left| \frac{f(z + h) - f(z)}{h} - \sum_{n=1}^{\infty} a_n n z^{n-1} \right|$$

$$= \left| \frac{1}{h} \sum_{n=1}^{\infty} a_n \left\{ (z + h)^n - z^n - hnz^{n-1} \right\} \right|$$

$$\leq \frac{1}{|h|} \sum_{n=1}^{\infty} |a_n| \left| \frac{h}{\delta} \right|^2 (|z| + \delta)^n$$

$$= \frac{|h|}{\delta^2} \sum_{n=1}^{\infty} |a_n|(|z| + \delta)^n \to 0 \ \text{as} \ h \to 0$$

The inequality follows from the binomial formula:

$$(z + h)^n - z^n - hnz^{n-1} = \sum_{k=0}^{n} \binom{n}{k} z^{n-k} h^k - z^n - hnz^{n-1}$$

$$= \sum_{k=2}^{n} \binom{n}{k} z^{n-k} h^k = \left\{ \sum_{k=2}^{n} \binom{n}{k} z^{n-k} h^{k-2} \right\} h^2$$

so

$$\left| (z+h)^n - z^n - hnz^{n-1} \right| \leq \sum_{k=2}^{n} \binom{n}{k} |z|^{n-k} |h|^{k-2} |h|^2$$

$$\leq \sum_{k=2}^{n} \binom{n}{k} |z|^{n-k} \delta^{k-2} |h|^2 = \sum_{k=2}^{n} \binom{n}{k} |z|^{n-k} \delta^k \left| \frac{h}{\delta} \right|^2$$

$$\leq \sum_{k=0}^{n} \binom{n}{k} |z|^{n-k} \delta^k \left| \frac{h}{\delta} \right|^2 = (|z| + \delta)^n \left| \frac{h}{\delta} \right|^2$$

$\square$

Definition 5.

Let Ω be an open subset of $\mathbf{C}$. A function $f : \Omega \to \mathbf{C}$ is said to be analytic *on Ω if for each point $z_0 \in \mathbf{C}$ there exists an open disc $B(z_0, \rho)$ in Ω in which f can be written as the sum of a power series centered at z_0, i.e. f can be written*

$$f(z) = \sum_{n=0}^{\infty} a_n (z - z_0)^n \ \text{for} \ z \in B(z_0, \rho)$$

We have already observed that the coefficients $\{a_n\}$ in the power series around z_0 are uniquely determined by f (Proposition I.6). By help of Proposition 4 we can even say what the coefficients are :

$$a_n = \frac{1}{n!} f^{(n)}(z_0) \ \text{for} \ n = 0, 1, 2, \cdots$$

So, restricting to the real line, we see that the power series expansion of an analytic function coincides with its Taylor series expansion. Given any sequence $\{a_n\}$ of complex numbers the power series $\sum_{n=0}^{\infty} a_n z^n$ may or may not converge around 0. There always exists a C^{∞} function on the real line with $\sum_{n=0}^{\infty} a_n z^n$ as its Taylor series. Many in fact (See e.g. [Me]). But such functions will in general not be restrictions of power series to the real line.

A consequence of Proposition 4 is that an analytic function is holomorphic, a result which is nice, but certainly not surprising. What is surprising and remarkable is that the converse is true (as will be proved in Chapter IV). Under the (apparently) humble assumption that f is complex differentiable, we shall infer that f is infinitely often differentiable and even that f around each point in its domain of definition can be expanded in a power series.

The situation is vastly different in the case of functions on the real line. There the derivative of a differentiable function need not even be continuous, let alone expandable in a power series.

But for functions in the complex plane the word "holomorphic" is synonymous with the word "analytic". However, until we have proved that result, we must distinguish between the two concepts.

Section 2 Line integrals

In this paragraph we introduce the line integrals which are important tools in complex function theory.

We shall until further notice not require any finer theory of integration and convergence theorems. All we need is to consider piecewise continuous functions on the real line and uniform convergence.

Definition 6.

A curve (i.e. a continuous map) $\gamma : [a, b] \to \mathbf{C}$, where $-\infty < a < b < \infty$, is said to be a path *if it is piecewise differentiable. This means that there are points*

$$a = t_0 < t_1 < t_2 < \cdots < t_{n-1} < t_n = b$$

such that the derivative γ' exists in the open subinterval $]t_j, t_{j+1}[$ and extends to a continuous function on the closed subinterval $[t_j, t_{j+1}]$ for each $j = 0, 1, 2, \cdots$.

The length *$l(\gamma)$ of the path γ is the number*

$$l(\gamma) := \int\limits_a^b |\gamma'(t)| dt \in [0, \infty[$$

To be precise, we distinguish between the curve γ, which is a map, and its image $\gamma^* := \gamma([a, b])$ which is a compact subset of $\mathbf{C}$.

We will say that a complex number z belongs (or does not belong) to the curve γ if $z \in \gamma^*$ (resp. $z \notin \gamma^*$). We express the same thing by saying that γ passes through z (resp. does not pass through z).

The curve γ is said to be closed *if $\gamma(a) = \gamma(b)$.*

If γ^* is contained in a subset Ω of $\mathbf{C}$ then we will say that γ is a curve in Ω.

Examples 7.

(a) The positively oriented circle γ around $z_0 \in \mathbf{C}$ with radius $R > 0$, i.e. the curve

$$\gamma(t) = z_0 + Re^{it} \text{ for } t \in [0, 2\pi]$$

is a closed path with length $2\pi R$. We will often write $|z - z_0| = R$ instead of γ.

(b) The line segment $[A, B]$, where A and B are complex numbers, i.e. the curve

$$[A, B] := (1 - t)A + tB \text{ for } t \in [0, 1]$$

is a path of length $l([A, B]) = |B - A|$.

(c) Let $A, B, C \in \mathbf{C}$. By the boundary $\partial\Delta$ of the triangle Δ with vertices A, B and C (the order is important!) we understand the closed curve defined by

$$\partial \Delta (t) := \begin{cases} [A, B](t) & \text{for } t \in [0, 1] \\ [B, C](t - 1) & \text{for } t \in [1, 2] \\ [C, A](t - 2) & \text{for } t \in [2, 3] \end{cases}$$

17

It is a path of length $l(\partial\Delta) = |B - A| + |C - B| + |A - C|$.

The length of a path does not depend on its parametrization. More formally :

Proposition 8.
Let $\gamma : [a, b] \to \mathbf{C}$ be a path. Let $\phi : [c, d] \to [a, b]$ be a continuously differentiable monotone map of $[c, d]$ onto $[a, b]$. Then $l(\gamma) = l(\gamma \circ \phi)$.

Proof :
The change of variables theorem.   □

Definition 9.
Let $\gamma : [a, b] \to \mathbf{C}$ be a path, and let $f : \gamma^* = \gamma([a, b]) \to \mathbf{C}$ be a continuous function. The line integral of f along γ is the complex number

$$\int_\gamma f = \int_\gamma f(z)dz := \int_{[a,b]} f(\gamma(t))\gamma'(t)dt$$

To emphasize what the variable is, we sometimes write $\int_\gamma f(w)dw$ or the like instead of $\int_\gamma f(z)dz$.

The next proposition collects properties of the line integral which will be useful for us in the sequel.

Proposition 10.
Let $\gamma : [a, b] \to \mathbf{C}$ be a path, and let f be a continuous, complex-valued function on γ^*. Then
(a)

$$\left| \int_\gamma f(z)dz \right| \leq \sup_{z \in \gamma^*} \{|f(z)|\}\, l(\gamma)$$

(b) (A primitive version of the fact that the line integral is invariant under orientation-preserving changes of parameter).
Let $-\infty < c < d < \infty$. If $\phi : [c, d] \to [a, b]$ has the form

$$\phi(s) = ks + l \ , \ s \in [c, d] \ , \ \text{where } k, l \in \mathbf{R}$$

and if ϕ maps $[c, d]$ onto $[a, b]$ then

$$\int_\gamma f = sign(k) \int_{\gamma \circ \phi} f$$

18

For two points A and B in **C** *we have in particular*

$$\int_{[A,B]} f = -\int_{[B,A]} f$$

(c) Let $c \in]a,b[$ and define paths γ_1 and γ_2 by

$$\gamma_1(t) := \gamma(t) \ \text{for} \ t \in [a,c]$$
$$\gamma_2(t) := \gamma(t) \ \text{for} \ t \in [c,b]$$

Then

$$\int_{\gamma} f = \int_{\gamma_1} f + \int_{\gamma_2} f$$

If $C \in [A,B]^*$, where A and B are points of **C**, *then in particular*

$$\int_{[A,B]} f = \int_{[A,C]} f + \int_{[C,B]} f$$

(d) If $\{f_n\}$ is a sequence of continuous functions on γ^* that converges uniformly on γ^* to f, then

$$\int_{\gamma} f_n \to \int_{\gamma} f \ \text{as} \ n \to \infty$$

(e) If f is defined on a neighborhood of γ^* and has a primitive, say F, there, then

$$\int_{\gamma} f = F(\gamma(b)) - F(\gamma(a))$$

If furthermore γ is closed then $\int_{\gamma} f = 0$.

Proof : Left to the reader. $\square$

We conclude this chapter by some applications of line integrals. They are investments for the future.

Definition 11.

A subset A of **R**^{n} *is said to be* starshaped, *if there exists a point $a \in A$ such that all the segments*

$$\{tx + (1-t)a | t \in [0,1]\} \ , \ x \in A$$

are contained in A; we say that A is starshaped *with respect to a.*

Clearly any convex set is starshaped. An interval is a simple, but important special case.

Lemma 12.

Let Ω be an open subset of $\mathbf{C}$ which is starshaped with respect to a point $a \in \Omega$. Let $g \in C(\Omega)$ have the property that $\int_{\partial\Delta} g = 0$, whenever Δ is a triangle in Ω with a as one of its vertices. Then the function

$$G(z) := \int_{[a,z]} g \ , \ z \in \Omega$$

is holomorphic on Ω and $G' = g$. In particular g has a primitive.

Proof: Let $z \in \Omega$. If $h \in \mathbf{C}$ is so small that $[z, z + h]$ is contained in Ω then we get from our assumption that

$$G(z + h) - G(z) = \int_{[a,z+h]} g - \int_{[a,z]} g = \int_{[z,z+h]} g = h \int_0^1 g(z + th)dt$$

Division through by h and use of the continuity of g at z gives us the lemma. $\qquad\square$

We can now give the promised simpler proof of Proposition 4:

Proof of Proposition 4: Let us, as in the earlier proof, for convenience take $z_0 = 0$. The power series

$$g(z) := \sum_{n=1}^{\infty} a_n n z^{n-1}$$

converges according to Corollary I.4 uniformly on compact subsets of $B(0, \rho)$, so term by term integration is permitted and yields

$$f(z) = \int_{[0,z]} g(w)dw + a_0 \ \text{ for } \ z \in B(0, \rho)$$

By Lemma 12 it now suffices to prove that the integral of g along any triangle Δ in $B(0, \rho)$ vanishes. But that is easy: Indeed,

$$\int_{\partial\Delta} g(z)dz = \sum_{n=1}^{\infty} a_n n \int_{\partial\Delta} z^{n-1} dz = 0$$

because the integral of z^{n-1} vanishes (z^{n-1} has a primitive, viz. z^n/n). $\qquad\square$

Lemma 13.

Let $\phi : \gamma^* \to \mathbf{C}$ be a continuous function on a path γ. Then the function

$$f(z) := \frac{1}{2\pi i} \int_\gamma \frac{\phi(w)}{w - z} dw \ , \ z \in \mathbf{C}\backslash\gamma^*$$

is analytic, hence holomorphic and

$$f^{(n)}(z) = \frac{n!}{2\pi i} \int_\gamma \frac{\phi(w)}{(w - z)^{n+1}} dw \ \text{ for } \ n = 0, 1, 2, \cdots$$

If $z_0 \in \mathbf{C}\backslash\gamma^*$ then the power series expansion of f around z_0 converges everywhere in the open disc around z_0 with radius $\text{dist}(z_0, \gamma^*)$.

As support for our memory we note that the formula for $f^{(n)}$ appears when we differentiate f under the integral sign.

Proof:

If $|z - z_0| < \text{dist}(z_0, \gamma^*)$, then the geometric series

$$\sum_{n=0}^{\infty} \left(\frac{z - z_0}{w - z_0} \right)^n$$

converges uniformly for $w \in \gamma^*$ to

$$\frac{1}{1 - \frac{z - z_0}{w - z_0}}$$

so writing

$$w - z = \left\{ 1 - \frac{z - z_0}{w - z_0} \right\} (w - z_0)$$

we find

$$2\pi f(z) = \int_\gamma \frac{\phi(w)}{w - z} dw = \int_\gamma \sum_{n=0}^{\infty} \left(\frac{z - z_0}{w - z_0} \right)^n \frac{\phi(w)}{w - z_0} dw$$

$$= \sum_{n=0}^{\infty} \int_\gamma \left(\frac{z - z_0}{w - z_0} \right)^n \frac{\phi(w)}{w - z_0} dw$$

$$= \sum_{n=0}^{\infty} \left\{ \int_\gamma \frac{\phi(w)}{(w - z_0)^{n+1}} dw \right\} (z - z_0)^n$$

which exhibits the power series expansion of f around z_0. $\qquad\square$

Section 3 Exercises

1. Let f be holomorphic on an open subset Ω of the complex plane. Define a function f^* on $\Omega^* := \{z \in \mathbf{C} | \bar{z} \in \Omega\}$ by $f^*(z) := \overline{f(\bar{z})}$ for $z \in \Omega^*$.

Show that f^* is holomorphic on Ω^*.

Suppose that f has the power series expansion $f(z) = \sum_{n=0}^{\infty} a_n z^n$ on Ω. Find the power series expansion of f^* on Ω^*.

2. Show that the function $f(z) := |z|^2$ has a complex derivative at $z = 0$ but nowhere else.

3. Is the function $f(z) = \Re(z)$ entire ?

4. Let u and v be real valued functions on an open subset Ω of $\mathbf{C}$. We assume that their first partial derivatives with respect to x and y exist and are continuous.

Show that $f := u + iv$ is holomorphic in Ω if the Cauchy-Riemann equations hold.

5. Show that

$$f(z) := \begin{cases} \exp\left(-1/z^4\right) & z \in \mathbf{C}\backslash\{0\} \\ 0 & z = 0 \end{cases}$$

is a discontinuous function, that nevertheless satisfies the Cauchy-Riemann equations everywhere.

6. Show that the pair of functions

$$u(x,y) = \frac{x}{x^2 + y^2} \quad \text{and} \quad v(x,y) = -\frac{y}{x^2 + y^2}$$

satisfies the Cauchy-Riemann equations on $\mathbf{C}\backslash\{0\}$.

Hint: You need not differentiate.

7. Let γ be a closed path in $\mathbf{C}$. Show that $\int_{\gamma} \bar{z}\,dz$ is purely imaginary.

8. Assume that the power series $f(z) = \sum_{n=0}^{\infty} a_n z^n$ converges in $\mathrm{B}(0,\rho)$. Show that f is analytic on $\mathrm{B}(0,\rho)$.

Hint:

$$z^n = \frac{1}{2\pi i} \int\limits_{|w|=R} \frac{w^n}{w - z}\,dw \quad \text{when } |z| < R < \rho$$

9. Let f be holomorphic on an open subset Ω of the complex plane. Let $\gamma :\,]-1,1[\to \Omega$ be differentiable at $t = 0$.

Show that $f \circ \gamma$ is differentiable at $t = 0$ and that

$$\frac{d(f \circ \gamma)}{dt}(0) = f'(\gamma(0))\gamma'(0)$$

Chapter 3 The Exponential Function, the Logarithm and the Winding Number

Section 1 The exponential function

We assume that the reader is familiar with the exponential function, the logarithm and the cos and sin functions on the real line.

We define the *exponential function* $\exp : \mathbf{C} \to \mathbf{C}$ by

$$\exp(z) := \sum_{n=0}^{\infty} \frac{z^n}{n!} \ \text{ for } \ z \in \mathbf{C}$$

By Theorem I.3(c) the power series converges absolutely and uniformly on compact subsets of $\mathbf{C}$, and by Proposition II.4 its sum $f(z) = \exp(z)$ is entire and satisfies $f' = f$, $f(0) = 1$. On the real line exp reduces to the well known real exponential function.

Let us study the exponential function on the imaginary axis in $\mathbf{C}$: For any $x \in \mathbf{R}$ we have

$$\exp(ix) = \sum_{n=0}^{\infty} \frac{(ix)^n}{n!} = \sum_{m=0}^{\infty} \frac{(ix)^{2m}}{(2m)!} + \sum_{m=0}^{\infty} \frac{(ix)^{2m+1}}{(2m+1)!}$$
$$= \sum_{m=0}^{\infty} (-1)^m \frac{x^{2m}}{(2m)!} + i \sum_{m=0}^{\infty} (-1)^m \frac{x^{2m+1}}{(2m+1)!}$$
$$= \cos x + i \sin x$$

From this we see in particular that $\phi(x) := \exp(ix)$ maps the real line onto the unit circle $\{ z \in \mathbf{C} \mid |z| = 1 \}$.

The most important property of the exponential function is that it couples the additive and multiplicative structures of $\mathbf{C}$ as follows :

Theorem 1 (The Addition Theorem).

$$\exp(a+b) = \exp(a)\exp(b) \ \text{ for all } \ a, b \in \mathbf{C}$$

Proof: When we differentiate $g(z) := \exp(z)\exp(a + b - z)$ we get $g'(z) = 0$, so g is constant. The contents of the theorem are $g(0) = g(a)$. □

Note that exp never vanishes (by The Addition Theorem).

The power series for exp has real coefficients. Therefore $\exp(\bar{a}) = \overline{\exp(a)}$, which by the addition theorem implies that

$$|\exp(a)|^2 = \exp(a)\exp(\bar{a}) = \exp(a + \bar{a}) = \exp(2\Re a) = \{\exp(\Re a)\}^2$$

23

so $|\exp(a)| = \exp(\Re a)$. From this we deduce that

$$|\exp(ia)| = 1 \iff a \in \mathbf{R}$$

We continue with a study of the exponential function on $\mathbf{C}$: Writing $w \in \mathbf{C}\backslash\{0\}$ in the form $w = |w|\frac{w}{|w|}$ we can by the above results express w as

$$w = \exp(x)\exp(i\theta) = \exp(x + i\theta) \text{ for some } x, \theta \in \mathbf{R}$$

so $\exp(\mathbf{C}) = \mathbf{C}\backslash\{0\}$. Finally,

$$\exp(a) = \exp(b) \iff a - b \in 2\pi i \mathbf{Z}$$

Proof :

$\Leftarrow$: An immediate consequence of The Addition Theorem.

$\Rightarrow$: Let $\exp a = \exp b$. Then $\exp(a - b) = 1$, and so $a - b \in i\mathbf{R}$, say $a - b = i\theta$. Then $\cos\theta + i\sin\theta = \exp i\theta = \exp(a - b) = 1$, so $\theta \in 2\pi\mathbf{Z}$. $\square$

Lemma 2.

Let $\phi(\theta) := \exp(i\theta)$ for $\theta \in \mathbf{R}$. Then ϕ is 1-1 on any half open interval of the form $[\alpha, \alpha + 2\pi[$, and it is a homeomorphism of the open interval $]\alpha, \alpha + 2\pi[$ onto the arc

$$\{z \in \mathbf{C} \mid |z| = 1, z \neq \exp(i\alpha)\}$$

A homeomorphism is a bijection which is continuous both ways.

The lemma expresses the geometrically obvious fact that the angle of a point depends continuously on the point.

Proof: The only non-trivial statement is the one about the homeomorphism property. ϕ is, however, clearly continuous, so left is just the continuity of the inverse of ϕ, i.e. the following claim:

$$\text{If } \exp(i\theta_n) \to \exp(i\theta) \text{ where } \theta_n, \theta \in]\alpha, \alpha + 2\pi[\text{ then } \theta_n \to \theta$$

To see this claim, let θ_0 be any limit point of the sequence $\{\theta_n\}$. θ_0 must belong to the closed interval $[\alpha, \alpha + 2\pi]$, and $\exp(i\theta) = \exp(i\theta_0)$. Now, θ_0 cannot equal any of the end points because exp there equals $\exp(i\alpha) \neq \exp(i\theta)$, and because exp is 1-1 on $]\alpha, \alpha + 2\pi[$, we must have $\theta = \theta_0$. $\square$

Section 2 Logarithm, argument and power

By §1 the restriction of the exponential function to $\mathbf{C}$ is the familiar real exponential function. Its inverse mapping is (by definition) the logarithm, which here will be denoted by Log. If $z \in \mathbf{C}\backslash\{0\}$ we can write

$$(1) \quad z = |z|\frac{z}{|z|} = \exp(Log\,|z|)\exp(i\theta) = \exp(Log\,|z| + i\theta)$$

with θ real. From §1 we see that θ is unique if we restrict it to lie in any given half-open interval of length 2π.

Definition 3. *Let* $z \in \mathbf{C} \backslash \{0\}$. *The sets*

$$(1) \quad \arg z := \left\{ \theta \in \mathbf{R} \mid \exp(i\theta) = \frac{z}{|z|} \right\} \text{ and}$$
$$(2) \quad \log z := \{ a \in \mathbf{C} \mid \exp(a) = z \}$$

are called the argument of z *and the logarithm of* z, *respectively. We will, however, use this terminology in an ambiguous way : Any real number in the set (2) will be called an argument of z and denoted by* $\arg z$; *any complex number in (3) will be called a logarithm of z and denoted by* $\log z$. *It will be clear from the context which interpretation we have in mind.*

The geometrical meaning of $\arg z$ is the following : Let l be the ray in $\mathbf{C}$ starting at the origin and passing through z. Then $\arg z$ is the set of values of the angles from the positive real semi-axis to the ray l.

We see from the definitions (1) and (2) that every non-zero complex number has an infinity of logarithms and arguments. Any two arguments differ by an integer multiple of 2π, and any two logarithms by an integer multiple of $2\pi i$. We can write

$$(4) \quad \log z = Log\,|z| + i \arg z \text{ for } z \in \mathbf{C} \backslash \{0\}$$

Definition 4.

If f is a never vanishing complex valued function, defined on some topological space, then by a continuous logarithm of f (or a branch of $\log f$) we mean any continuous complex valued function F such that $\exp F = f$, and by a continuous argument of f (or a branch of $\arg f$) we mean any continuous real valued function θ such that $f = |f| \exp(i\theta)$.

Of course, if $f : X \to\,]0, \infty[$ is continuous, then $Log\, f$ is an example of a branch of $\log f$.

From the discussion above we get a uniqueness result : If f is a never-vanishing continuous complex valued function on a connected topological space, then any two continuous logarithms of f differ by a constant integer multiple of $2\pi i$; and any two continuous arguments of f differ by a constant integer multiple of 2π.

We now turn to the question of regularity of logarithms and arguments. The following theorem tells in particular that any branch of $\log z$ automatically is holomorphic. But more is true: In Chapter IV we shall present a generalization in which we replace exp by any non-constant holomorphic function (Proposition IV.18).

Theorem 5.

Let f be holomorphic on an open subset Ω of the complex plane, and let F be a continuous logarithm of f in Ω.

Then F is also holomorphic in Ω, and $F' = f'/f$.

Proof:

We shall show that F has a complex derivative at each point $z \in \Omega$. From the defining relation $\exp F(a) = f(a)$ for any $a \in \Omega$ we get

$$\frac{\exp\left(F(z+h)\right) - \exp F(z)}{h} \to f'(z) \quad \text{as} \quad h \to 0$$

and division through by $\exp F(z) = f(z)$ yields

$$(5) \quad \frac{\exp\left(F(z+h) - F(z)\right) - 1}{h} \to \frac{f'(z)}{f(z)} \quad \text{as} \quad h \to 0$$

Expansion of the exponential function in (5) shows that

$$\frac{F(z+h) - F(z)}{h} + \frac{1}{h} \sum_{n=2}^{\infty} \frac{\{F(z+h) - F(z)\}^n}{n!} \to \frac{f'(z)}{f(z)} \quad \text{as} \quad h \to 0$$

so it suffices to prove that

$$(6) \quad \frac{1}{|h|} \sum_{n=2}^{\infty} \frac{|F(z+h) - F(z)|^n}{n!} \to 0 \quad \text{for} \quad h \to 0$$

Now $|e^a - 1| \geq |a|/2$ when $|a|$ is small. Indeed,

$$\frac{e^a - 1}{a} \to 1 \quad \text{as} \quad a \to 0$$

Therefore we find, putting $a = F(z+h) - F(z)$ in (5), that $|F(z+h) - F(z)|/|h|$ is bounded for small h, i.e. there exist $C > 0$ and $\delta \in \,]0,1[$ such that

$$|h| < \delta \quad \Rightarrow \quad |F(z+h) - F(z)| \leq C|h|$$

By help of this we may for $|h| < \delta$ estimate the left hand side of (6) as follows:

$$\frac{1}{|h|} \sum_{n=2}^{\infty} \frac{|F(z+h) - F(z)|^n}{n!} \leq \frac{1}{|h|} \sum_{n=2}^{\infty} C^n |h|^n$$

$$\leq |h| \sum_{n=2}^{\infty} \frac{C^n}{n!} \leq |h| e^C \to 0 \quad \text{as} \quad h \to 0$$

$\square$

Example 6.

Consider, for given $\theta_0 \in \mathbf{R}$, the open set

$$\Omega := \left\{ z \in \mathbf{C} \backslash \{0\} \mid \frac{z}{|z|} \neq \exp(i\theta_0) \right\}$$

Let $\theta(z)$ for $z \in \Omega$ denote the value of $\arg z$ in the open interval $]\theta_0, \theta_0 + 2\pi[$. Then

(α) $z \to \theta(z)$ is a branch of $\arg z$ in Ω.

(β) $z \to Log\,|z| + i\theta(z)$ is a branch of $\log z$ in Ω. In particular it is a holomorphic function.

Proof:

(α) is a consequence of the last statement of Lemma 2, while (β) follows from (α). □

Example 7.

Let us take $\theta_0 = -\pi$ in Example 6, so that $\Omega = \mathbf{C} \backslash \{z \in \mathbf{R} \mid z \leq 0\}$. Let Arg z for $z \in \Omega$ denote the value of $\arg z$ in the interval $]-\pi, \pi[$. Then $z \to Arg\,z$ is a branch of $\arg z$ in Ω. It is called the *principal branch of* $\arg z$. The corresponding continuous logarithm of z in Ω is called the *principal branch of* $\log z$ and denoted Log z . So

$$Log\,z = Log\,|z| + iArg\,z \ \text{ for } \ z \in \Omega$$

Log z is holomorphic according to Theorem 5.

Note that the principal branch of $\log z$ extends the function Log on $\mathbf{R}$ from earlier so that the notation is unambiguous. If $x \in \mathbf{R}^+$ and nothing is specified to the contrary, we will very often be sloppy and write $\log x$ instead of the more precise term $Log\,x$. Note also that $(Log\,z)' = z^{-1}$.

Now we can define complex powers : Let $a \neq 0$. For any *fixed* choice of $\log a$ - and we have a countable infinity of choices -

$$f(z) := \exp(z \log a)$$

is called *the* z^{th} *power of* a. f is clearly holomorphic in the entire complex plane, i.e. f is entire. The Addition Theorem tells us that f satisfies the relations

$$f(z + w) = f(z)f(w)\,, \ f(0) = 1\,, \ f(1) = a$$

In particular

$$f(n) = a^n \ \text{ for } \ n = 0, 1, 2, \cdots$$

so it is natural to use the notation a^z for $f(z)$. Note also that a^n is independent of the choice of $\log a$. We see that $\exp z$ is one of the zth powers of e, viz the one with the choice $\log e = 1$, and we shall from now on often write e^z instead of $\exp z$. More generally, if $a > 0$ we shall always put

$$a^z = \exp(z \, Log \, a) \text{ for } z \in \mathbf{C}$$

Section 3 Existence of continuous logarithms

In this section we shall discuss whether branches of $\log f$ exist for given function f. Before studying general functions f we will examine the important special case $f(z) = z$. To indicate the kind of difficulties we may encounter, we present an example to the effect that $\log z$ cannot be defined in a continuous way in all of $\mathbf{C}\backslash\{0\}$. However, to make the reader happy again, we note that there do exist branches of $\log z$ in the complex plane minus any half-ray starting at the origin (Example 6 above).

Example 8.

We claim that it is not possible to find a continuous $\log z$ on

$$S^1 := \{z \in \mathbf{C} \mid |z| = 1\}$$

let alone on the punctured plane $\mathbf{C}\backslash\{0\}$. It suffices to prove that there does not exist any branch of $\arg z$ on S^1 - after all $\arg z$ is the imaginary part of $\log z$. It is intuitively obvious that no branch of $\arg z$ exists on S^1 : Let us keep track of the argument as we go around the unit circle counterclockwise starting at 1. From the geometrical interpretation of the argument we see that $\arg z$ increases continuously, and when we have gone all the way round it has increased by 2π. And so it does not take its former value at 1 as it should by its continuity at 1. Thus $\arg z$ cannot be defined in a continuous way all around the unit circle.

Here is a short rigorous proof of the non-existence of $\arg z$: As most proofs of non-existence it goes by contradiction, so we assume that there exists a continuous argument $\theta : S^1 \to \mathbf{R}$, so that

$$z = \exp(i\theta(z)) \text{ for all } z \in S^1$$

The relation shows in particular that θ is 1-1. Thus the mapping $f : S^1 \to \{1, -1\}$, given by

$$f(z) := \frac{\theta(z) - \theta(-z)}{|\theta(z) - \theta(-z)|} \text{ for } z \in S^1$$

is well-defined and continuous. But since $f(z) = -f(-z)$ it maps the connected space S^1 onto the disconnected space $\{1, -1\}$. Contradiction ! □

We shall now treat the question of whether a given function f has a continuous logarithm $\log f$, i.e. the existence question. We hide most of the technicalities away in the proof of the following proposition.

28

Proposition 9 (The homotopy lifting property).

Let X be a topological space and let $F : X \times [0,1] \to \mathbf{C}\backslash\{0\}$ be a continuous map. Assume that the restriction of F to $X \times \{0\}$ has a continuous logarithm, say $\phi : X \times \{0\} \to \mathbf{C}$.

Then F has a continuous logarithm that extends ϕ.

Proof : We use the abbreviation $\phi(x) = \phi(x,0)$ for $x \in X$.

1st step.

We will prove that it suffices to verify the following localized version of the proposition :

To every $x \in X$ there exists an open neighborhood $N(x)$ of x in X, and a continuous map

$$\theta_x : N(x) \times [0,1] \to \mathbf{C} \text{ , such that}$$
$$\theta_x(y,0) = \phi(y) \text{ for all } y \in N(x) \text{ , and}$$
$$\exp(\theta_x) = F \text{ on } N(x) \times [0,1]$$

Indeed, assume that the localized version is true. Then for any $x, y \in X$ and any $x_0 \in N(x) \cap N(y)$ we have

$$\exp(\theta_x(x_0,t)) = F(x_0,t) = \exp(\theta_y(x_0,t)) \text{ so}$$
$$\theta_x(x_0,t) - \theta_y(x_0,t) = 2\pi i\, n(t) \text{ , where } n(t) \in \mathbf{Z}$$

The left side is a continuous function of t, so $t \to n(t)$ is a continuous, integer valued function on [0,1] , hence a constant. Taking the condition

$$\theta_x(x_0,0) = \phi(x_0) = \theta_y(x_0,0)$$

into account we see that the constant is 0.

We have thus shown that θ_x and θ_y agree on their common domain of definition, and so the collection $\theta_x, x \in X$ patches together to a continuous function $\theta : X \times [0,1] \to \mathbf{C}$, which is the desired logarithm of F.

2nd step.

We shall verify the localized version of the proposition is true, so let $x_0 \in X$. By the compactness of [0,1] we determine points $0 = t_0 < t_1 < \cdots < t_n = 1$ and a neighborhood $N = N(x_0)$ of $x_0 \in X$ such that each

$$F(N \times [t_k, t_{k+1}]) \text{ , } k = 0, 1, 2, \cdots, n-1$$

is contained in a subset of $\mathbf{C} \setminus \{0\}$ on which there exists a branch of $\log z$.

We now define inductively continuous maps

$$\theta_k : N \times [t_k, t_{k+1}] \to \mathbf{C} \text{ for } k = 0, 1, 2, \cdots, n-1$$

in the following way :

Since there exists a branch of $\log z$ on $F(N \times [0, t_1])$ we can form $\log F$ there, i.e. we can find a continuous function $\Lambda_0 : N \times [0, t_1] \to \mathbf{C}$ such that

$$\exp(\Lambda_0) = F \text{ on } N \times [0, t_1]$$

Noting that

$$\exp(\Lambda_0(x, 0)) = F(x, 0) = \exp(\phi[x]) \text{ so that}$$
$$\exp[\phi(x) - \Lambda_0(x, 0)] = 1$$

we define

$$\theta_0(x, t) := \Lambda_0(x, t) - \Lambda_0(x, t) + \phi(x)$$

Then $\theta_0 : N \times [0, t_1] \to \mathbf{C}$ is a continuous function that satisfies

$$\exp(\theta_0) = F \text{ , and } \theta_0(x, 0) = \phi(x) \text{ for } x \in N$$

Arguing in exactly the same way we find successively for $k = 1, 2, \cdots, n-1$ continuous functions $\theta_k : N \times [t_k, t_{k+1}] \to \mathbf{C}$ such that

$$\exp(\theta_k) = F \text{ , and } \theta_k(x, t_k) = \theta_{k-1}(x, t_k) \text{ for } x \in N$$

Obviously the θ_k glue together to form a continuous function $\theta : N \times [0, 1] \to \mathbf{C}$ with the desired properties. $\qquad\square$

Remark. Proposition 9 is a special case of a result from point set topology, viz that a covering projection is a fibration. See e.g. [Sp;Theorem 2.2.3 p.67].

Proposition 10.
If A is a starshaped subset of $\mathbf{R}^n$ then any continuous mapping $f : A \to \mathbf{C} \backslash \{0\}$ has a continuous logarithm.

Proof: With notation from Definition II.11 we consider the continuous map $F : A \times [0, 1] \to \mathbf{C} \backslash \{0\}$, defined by

$$F(x, t) := f(tx + (1 - t)a) \text{ for } (x, t) \in A \times [0, 1]$$

Since $x \to F(x, 0) = f(a)$ is a constant function it has a continuous logarithm, hence by Proposition 9 so does F. A fortiori so does $x \to F(x, 1) = f(x)$. $\qquad\square$

As we shall see later (Theorem V.9), Proposition 10 extends to from starshaped to simply connected spaces.

Here we will as an application of Proposition 10 present an interesting digression: The 2-dimensional version of the famous Brouwer's fixed point theorem (Theorem 11(c) below).

Theorem 11.

(a) There is no continuous mapping $r : B[0,1] \to S^1$ with the property that $r(z) = z$ for all $z \in S^1$.

(b) Any continuous mapping $f : B[0,1] \to B[0,1]$ has a fixed point.

(c) Let K be a non-empty, convex, compact subset of $\mathbf{C}$. Then any continuous map $f : K \to K$ has a fixed point.

Proof :

(a) Let us assume that such a map r exists. $B[0,1]$ is starshaped, so we can by Theorem 10 find a continuous function $R : B[0,1] \to \mathbf{C}$ such that $\exp(R) = r$. Restriction to S^1 yields that $\exp(R(z)) = z$ for all $z \in S^1$, so R is a branch of $\log z$ on S^1. But such one does not exist (by Example 8), so we have arrived at a contradiction.

(b) If f has no fixed point then we can construct a map r as in (a) as follows: Let $r(x)$ be the intersection point of S^1 with the ray from $f(x)$ through x.

(c) We may without loss of generality assume that $K \subseteq B[0,1]$. Let $\phi : B[0,1] \to K$ be the mapping that to $z \in B[0,1]$ associates the nearest point to z in K. By (b) the mapping $f \circ \phi$ has a fixed point. And that point necessarily belongs to K, because the image of f is contained in K. □

Section 4 The winding number

In this paragraph we introduce the notion of winding number (= index) which will turn out to be of central importance.

Definition 12.

Let $\gamma : [a, b] \to \mathbf{C}$ be a closed curve. If $z \in \mathbf{C} \backslash \gamma^*$ then the winding number *or* index *of γ relative to z is the integer*

$$Ind_\gamma(z) := \frac{\log(\gamma(b) - z) - \log(\gamma(a) - z)}{2\pi i}$$

where $t \to \log(\gamma(t) - z)$ is any continuous logarithm of $t \to \gamma(t) - z$.

Pertinent remarks.

(1) By Proposition 10 there exists a branch of $\log(\gamma(t) - z)$.

(2) Since any two continuous logarithms differ by a constant, we see that $Ind_\gamma(z)$ does not depend on our choice of branch of logarithm.

(3) The reason why $Ind_\gamma(z)$ is called the winding number is the following: Let us write

$$\log(\gamma(t) - z) = Log|\gamma(t) - z| + i\theta(t)$$

where $t \to \theta(t) \in \arg(\gamma(t) - z)$ is a continuous argument of $t \to \gamma(t) - z$. Then

$$Ind_\gamma(z) = \frac{\theta(b) - \theta(a)}{2\pi}$$

Now, $\theta(t)$ is geometrically the angle between the positive half of the x-axis and the ray from z through $\gamma(t)$, so $Ind_\gamma(z)$ measures the net increase in that angle as the parameter t ranges from a to b. Thus $Ind_\gamma(z)$ is the number of times (counted with sign) the ray from z through $\gamma(t)$ performs a full rotation (in the counterclockwise direction) as t increases from a to b, i.e. how often the curve γ winds around the point z.

(4) The concept of winding number of a closed curve has been generalized to higher dimensions, where it is replaced by the concept of degree of a map.

Example 13.

Let $\triangle$ be the triangle with vertices 1, $\exp(2\pi i/3)$ and $\exp(4\pi i/3)$. We will compute the winding number of its boundary $\partial\triangle$ (defined in Example II.7(c)) relative to the origin 0.

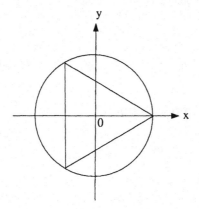

Note that $t \to Arg\,\partial\,\triangle\,(t)$ is a continuous $\arg z$-function along the two sides $\partial\,\triangle\,|_{[0,1]}$ and $\partial\,\triangle\,|_{[2,3]}$, and $t \to Arg_1\partial\,\triangle\,(t)$ is a continuous $\arg z$-function along the remaining side, where we let Arg_1 denote that branch of $\arg z$ that is defined in $\mathbb{C}\backslash[0,\infty[$ and takes values in $]0,2\pi[$. Now

$$\theta(t) := \begin{cases} Arg\,\partial\,\triangle\,(t) & \text{for } 0 \leq t \leq 1 \\ Arg_1\,\partial\,\triangle\,(t) & \text{for } 1 \leq t \leq 2 \\ Arg\,\partial\,\triangle\,(t) + 2\pi & \text{for } 2 \leq t \leq 3 \end{cases}$$

is a continuous $\arg z$-function along $\partial\triangle$, so $Ind_{\partial\triangle}(0) = 1$.

Example 14.

Consider the positively oriented circle around $z_0 \in \mathbb{C}$ with radius R, i.e. the closed curve $\gamma : [0, 2\pi] \to \mathbb{C}$, defined by

$$\gamma(t) = z_0 + Re^{it}$$

32

It is easy to compute the winding number of γ relative to z_0 : Indeed, $t \to Log\, R + it$ is a continuous logarithm along $\gamma(t) - z_0$, so by definition

$$Ind_\gamma(z_0) = \frac{(Log\, R + i2\pi) - (Log\, R + i0)}{2\pi i} = 1$$

Example 15.

The three curves $\gamma_1, \gamma_{-1}, \gamma_2 : [0, 2\pi] \to \mathbf{C} \backslash \{0\}$, given by

$$\gamma_j(t) := \exp(ijt) \text{ for } j = 1, -1, 2$$

have the same image, viz the unit circle, but nevertheless different winding numbers relative to 0, viz 1, -1 and 2. The example shows in particular that the index of a curve depends not just on the image γ^* of the curve γ, but also on how the moving point $\gamma(t)$ traces the image γ^*.

A trivial example: A constant curve has index 0.

We have in the above examples pedantically computed the index analytically, although the results are geometrically evident. From now on we will not bother to do so. If a closed curve has so simple a look that it is geometrically obvious what its index is, then we will not elaborate on the matter.

Before we list properties of the winding number we call the attention to a topological concept that will be handy now and later:

Definition 16.

Let Ω be a topological space.

(α) Two closed curves $\gamma_0, \gamma_2 : [a, b] \to \Omega$ are said to be homotopic *in Ω, if there exists a continuous map (a* homotopy*) $\Gamma : [a, b] \times [0, 1] \to \Omega$ such that*

$$\Gamma(t, 0) = \gamma_0(t) \text{ and } \Gamma(t, 1) = \gamma_1(t) \text{ for all } t \in [a, b] \text{ and}$$
$$\Gamma(a, s) = \Gamma(b, s) \text{ for all } s \in [0, 1]$$

(β) A closed curve is said to be null-homotopic, *if it is homotopic to a constant curve.*

The geometrical contents of (α) are that the curve γ_0 continuously is deformed into γ_1 as the parameter s increases from 0 to 1. The curves $t \to \Gamma(t, s)$ are closed curves for each fixed $s \in [0, 1]$ and they form intermediate steps in the deformation of γ_0 into γ_1.

Theorem 17 (Properties of the winding number).

(a) The winding number is independent of the parametrization of the basic interval as long as the orientation is unchanged: More precisely, if $\phi : [c, d] \to [a, b]$ is continuous and $\phi(c) = a$ and $\phi(d) = b$, then

$$Ind_{\gamma \circ \phi}(z) = Ind_\gamma(z) \text{ for all } z \in \mathbf{C} \backslash \gamma^*$$

(b) The winding number is a homotopy invariant, i.e. if two closed curves γ and τ are homotopic in $\mathbf{C}\backslash\{z_0\}$ then $Ind_\gamma(z_0) = Ind_\tau(z_0)$.

(c) Let γ be a closed curve in $\mathbf{C}$. Then the mapping $z \to Ind_\gamma(z)$ of $\mathbf{C}\backslash\gamma^*$ into $\mathbf{Z}$ is constant on each connected component of $\mathbf{C}\backslash\gamma^*$, and it is 0 on the unbounded component.

(d) Let γ be a closed path that does not pass through $z \in \mathbf{C}$. Then we have the following formula for the winding number :

$$Ind_\gamma(z) = \frac{1}{2\pi i} \int_\gamma \frac{dw}{w - z}$$

Proof:

(a) This is left to the reader.

(b) Let $\Gamma : [a, b] \times [0, 1] \to \mathbf{C}\backslash\{z_0\}$ be a homotopy between γ and τ. The function $(t, s) \to \Gamma(t, s) - z_0$ has a continuous logarithm F since its domain of definition is starshaped. Now the function

$$s \to \frac{F(b, s) - F(a, s)}{2\pi i} = Ind_{\Gamma(\cdot, s)}(z_0)$$

is a continuous, integer-valued function on [0,1] , hence a constant.

(c) Let $\gamma : [a, b] \to \mathbf{C}$. Let $z_0 \in \mathbf{C}\backslash\gamma^*$ and let $r := dist(z_0, \gamma^*) > 0$. The continuous mapping $f : B(z_0, r) \times [a, b] \to \mathbf{C}\backslash\{0\}$ given by $f(z, t) := \gamma(t) - z$ has a continuous logarithm F since its domain of definition is starshaped. The formula

$$Ind_\gamma(z) = \frac{F(z, b) - F(z, a)}{2\pi i} \text{ for } z \in B(z_0, r)$$

reveals that $Ind_\gamma(z)$ is a continuous function of z. Being also integer-valued it is constant on the connected components of $\mathbf{C}\backslash\gamma^*$.

It is left to prove that $Ind_\gamma(z_0) = 0$ for just one point z_0 in the unbounded component of $\mathbf{C}\backslash\gamma^*$, say the point $z_0 = 1 + \sup\{|\gamma(t)| \,|\, t \in [a, b]\}$. The homotopy

$$\Gamma(t, s) := (1 - s)\gamma(t) \text{ for } (t, s) \in [a, b] \times [0, 1]$$

contracts γ to the constant curve $\tau = 0$, so by homotopy invariance

$$Ind_\gamma(z_0) = Ind_\tau(z_0) = 0$$

(d) If h is a continuous logarithm along $\gamma(t) - z$, i.e. if $\exp(h(t)) = \gamma(t) - z$ then we find formally by differentiation that

$$h'(t) = \frac{\gamma'(t)}{\gamma(t) - z}$$

which tempts us to define the function h by

$$h(t) := \int_a^t \frac{\gamma'(s)}{\gamma(s) - z} ds + c$$

where the constant $c \in \mathbf{C}$ for later purposes is chosen so that $\exp(c) = \gamma(a) - z$. Note that h is continuous in [a,b] and that h is differentiable wherever γ is. An easy calculation reveals that the continuous function

$$\phi(t) := (\gamma(t) - z)e^{-h(t)}$$

has a vanishing derivative at each point where $\gamma'(t)$ exists, so ϕ is a constant. Since $\phi(a) = 1$ by our choice of c we have thus

$$e^{h(t)} = \gamma(t) - z \quad \text{for all} \ \ t \in [a, b]$$

so h is a continuous logarithm along $\gamma(t) - z$. Finally

$$Ind_\gamma(z) = \frac{h(b) - h(a)}{2\pi i} = \frac{1}{2\pi i} \int_\gamma \frac{dw}{w - z}$$

$\square$

Section 5 Square roots

An important concept, related to the concept of logarithm, is that of nth root. We concentrate on the special case of square roots and leave the generalization to n^{th} roots to the reader.

Definition 18.

Let $f : X \to \mathbf{C}$ be a complex-valued function on a topological space X. By a continuous square root of f (or a branch of $\sqrt{f}$) we mean any complex-valued continuous function $r : X \to \mathbf{C}$ satisfying $r^2 = f$.

An example is the standard square root function $\sqrt{\cdot} : [0, \infty[\to \mathbf{R}$ with the usual convention that $\sqrt{x}$ is the unique *positive* number $r(x)$ satisfying $r(x)^2 = x$. Of course $x \to -\sqrt{x}$ is another example of a continuous square root.

The uniqueness of square roots is the topic for the next proposition. Then comes regularity and finally existence.

Proposition 19.

Let r_1 and r_2 be two continuous square roots of a never vanishing function, all defined on a connected topological space.

35

Then either $r_1 = r_2$ or $r_1 = -r_2$. In particular, if r_1 and r_2 agree at one point they agree everywhere.

Proof:

$$\left(\frac{r_1}{r_2}\right)^2 = 1 \quad \text{so} \quad \frac{r_1(x)}{r_2(x)} \in \{-1, 1\}$$

Thus r_1/r_2 is a continuous integer-valued function on a connected space, hence a constant, i.e. either $+1$ or -1. □

Proposition 20.

Let $f : \Omega \to \mathbf{C}\backslash\{0\}$ be a holomorphic function defined on an open subset Ω of the complex plane. Then any continuous square root of f is holomorphic on Ω.

Remark. It is actually superfluous to assume that f never vanishes. See the remarks prior to Theorem IV.18.

Proof:

Let r be any continuous square root of f in Ω. Let B be any open disc in Ω. Now f has a holomorphic logarithm F in B (Theorems 10 and 5), and so $\exp(F/2)$ is a holomorphic square root of f in B. But $r|_B$ is by Proposition 19 either $\exp(F/2)$ or $\exp(-F/2)$. □

Proposition 21.

Any holomorphic, never vanishing function on a starshaped open subset of $\mathbf{C}$ has a holomorphic square root.

Proof:

Such a function has according to Proposition 10 a continuous logarithm F. A continuous square root is provided by $\exp(F/2)$, and that square root is holomorphic by Proposition 20. □

As we shall see later (Theorem V.9), Proposition 21 extends from starshaped to simply connected sets.

Examples 22.

(a) The function z has on $\mathbf{C}\backslash\,]-\infty, 0]$ the continuous square root

$$\sqrt{z} = \exp\left(\frac{Log\,|z| + i\,Arg\,z}{2}\right) = \sqrt{|z|}\,e^{i\frac{Arg\,z}{2}}$$

which on the positive axis reduces to the ordinary square root. We call this square root for the *principal square root of* z. It is holomorphic according to Proposition 20.

(b) We claim that there is no continuous square root of z on all of $\mathbf{C}\backslash\{0\}$, not even on S^1. Indeed, let $\sqrt{z}$ be one, and consider the curve $\gamma(t) := \sqrt{\exp(2\pi i t)}$. It has the property that $\gamma^2 = \tau$, where

$$\tau(t) = e^{2\pi i t} \text{ for } t \in [0,1]$$

Now $2\,Ind_\gamma(0) = Ind_\tau(0) = 1$, which contradicts the fact that the index is integer valued.

(c) As a third example we will study the function $z^2 - 1$ and its holomorphic square roots which enter naturally in many considerations.

Consider for example $\arcsin z$. This is a multiple valued function defined by $w = \arcsin z \;\Leftrightarrow\; \sin w = z$. On any open subset of the complex plane where a holomorphic function w, satisfying $\sin w(z) = z$, can be defined, we must have

$$w'(z) = \frac{1}{\cos w(z)} = \frac{1}{\sqrt{1-z^2}}$$

So a look at the holomorphic square root of $z^2 - 1$ will not be a waste of time.

Our result is that the function $z^2 - 1$ on $\mathbf{C}\backslash[-1,1]$ has exactly one holomorphic square root which is positive for $z \in\,]1,\infty[$.

Proof of that: It suffices to exhibit a continuous $\sqrt{z^2 - 1}$ on $\mathbf{C}\backslash[-1,1]$, because the statements about uniqueness and holomorphy are consequences of the Propositions 19 and 20. Let $\sqrt{z}$ be the principal square root of z on $\mathbf{C}\backslash\,] - \infty, 0]$ (described in (a) above). If $z \in \mathbf{C}\backslash[-1,1]$, then by a small computation

$$1 - z^{-2} \in \mathbf{C}\backslash\,] - \infty, 0]$$

so we can form the composite map $\sqrt{1 - z^{-2}}$. The function

$$r(z) := z\sqrt{1 - z^{-2}} \text{ for } z \in \mathbf{C}\backslash[-1,1]$$

is the desired continuous square root of $z^2 - 1$ on $\mathbf{C}\backslash[-1,1]$. $\qquad\Box$

Section 6 Exercises

1. Show that exp maps $\{z \in \mathbf{C}|-\pi < \Im z < \pi\}$ onto $\mathbf{C}\backslash\,] - \infty, 0]$ bijectively, and that the inverse map is Log.

Determine the images by exp of lines (resp. line segments) , parallel to the real (resp. imaginary axis).

2. Find the image by Log of $\{z \in \mathbf{C}\backslash\{0\} \,|\, |z - \tfrac{1}{2}| = \tfrac{1}{2}\}$.

3. Let log be that branch of $\log z$ in the snake shaped domain on the drawing below which has $\log 1 = 0$. What is $\log e$?

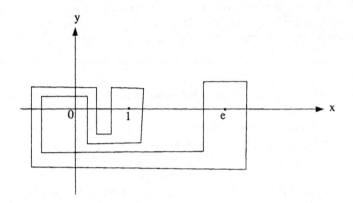

4. A topological space X is said to be *contractible* if there is a continuous map $c : X \times [0,1] \to X$ and a point $x_0 \in X$ such that

$$c(x,0) = x \ \text{ and } \ c(x,1) = x_0 \ \text{ for all } \ x \in X$$

Show that the circle S^1 is not contractive.

5. Show that

$$Log\,(1+z) = \sum_{n=1}^{\infty} \frac{(-1)^{n-1}}{n} z^n \ \text{ for } \ |z| < 1$$

Hint: Theorem 5.

6. Show that

$$\left(1 + \frac{z}{n}\right)^n \to \exp\,(z) \ \text{ as } \ n \to \infty$$

uniformly for z in any compact subset of **C**.

7. (a) (J. Bernoulli's paradox)

Is anything wrong in the following pretty argument: Taking Logs in the identity $(-z)^2 = z^2$ we get $2\,Log\,z = 2\,Log\,(-z)$, so $Log\,z = Log\,(-z)$.

(b) Compute $Log\,i$ and $Log(-i)$.

(c) Is the formula $Log\,(z_1 z_2) = Log\,(z_1) + Log\,(z_2)$ correct ?

8. Let $a_1, a_2, \cdots, a_n$ be complex numbers such that

$$|a_j| < 1 \ \text{ and } \ \left| \prod_{k=1}^{j} (1 + a_k) - 1 \right| < 1 \ \text{ for } \ j = 1, 2, \cdots, n$$

Show that

$$Log \left\{ \prod_{k=1}^{n} (1 + a_k) \right\} = \sum_{k=1}^{n} Log\,(1 + a_k)$$

38

9. Find the values of i^i.

10. Does there exist a branch of $\log z$ in

(α) $\{z \in \mathbf{C} \mid 1 < |z| < 2\}$?

(β) $\{z \in \mathbf{C} \mid \Re z > \Im z\}$?

11. Let $f, g : \mathbf{C} \to \mathbf{C}$ be continuous functions satisfying that $f^2 + g^2 = 1$. Show that there exists a continuous function $\phi : \mathbf{C} \to \mathbf{C}$ such that $f = \cos \phi$ and $g = \sin \phi$. Show that ϕ is unique modulo integer multiples of 2π, and that ϕ is holomorphic if f and g are holomorphic.

Hint: $f^2 + g^2 = (f + ig)(f - ig)$.

12. Let $\gamma, \rho : [0, 1] \to \mathbf{C} \backslash \{0\}$ be closed curves. Show that

(α) $Ind_{\gamma\rho}(0) = Ind_\gamma(0) + Ind_\rho(0)$

(β) $Ind_{-\gamma}(0) = -Ind_\gamma(0)$, where $(-\gamma)(t) = \gamma(1 - t)$

(γ) $Ind_{\gamma+\rho}(0) = Ind_\gamma(0) + Ind_\rho(0)$, where

$$(\gamma + \rho)(t) = \begin{cases} \gamma(t) & if \ t \in [0, 1] \\ \rho(t - 1) & if \ t \in [1, 2] \end{cases}$$

13. Let $\gamma : [0, 2\pi] \to \mathbf{C} \backslash \{0\}$ be the closed curve $\gamma(t) := 3 \cos t + i \sin t$. Show that $Ind_\gamma(0) = 1$.

14. Let A be a starshaped subset of $\mathbf{C}$. Let γ be a closed curve in A and let $z \in \mathbf{C} \backslash A$. Show that $Ind_\gamma(z) = 0$.

15. Let $A : \mathbf{C} \to \mathbf{C}$ be an **R**-linear isomorphism and let γ be a closed curve in $\mathbf{C} \backslash \{0\}$. Show that $Ind_{A \circ \gamma}(0) = sign(\det A) \, Ind_\gamma(0)$.

16. A merry man walks with his dog on lead 10 times around a lamp post. He never allows the dog so much leash that it can get around to the other side of the lamp post of the man. Assuming that they both return to their initial positions after the man's 10 rotations, show that also the dog has been 10 times around the lamp post.

Hint: The problem is a special case of the following mathematical problem: Let $\gamma, \sigma : [a, b] \to \mathbf{C} \backslash \{0\}$ be two closed curves satisfying

$$|\gamma(t) - \sigma(t)| \leq |\gamma(t)| \ \text{for all} \ t \in [a, b]$$

Show that $Ind_\gamma(0) = Ind_\sigma(0)$ (note for example that γ and σ are homotopic).

17. (Extension of Exercise 16). Let $\gamma, \sigma : [a, b] \to \mathbf{C}$ be closed curves satisfying that

$$|\gamma(t) - \sigma(t)| < |\gamma(t)| + |\sigma(t)| \ \text{for all} \ t \in [a, b]$$

Show first that neither γ nor σ passes through 0, and then that $Ind_\gamma(0) = Ind_\sigma(0)$. Hint : $\gamma(t)/\sigma(t)$ never takes values in $] - \infty, 0]$, so we may form $Log(\gamma/\sigma)$.

18. Prove the *Fundamental Theorem of Algebra* :

Any complex polynomial $p(z) = z^n + a_{n-1} z^{n-1} + \cdots + a_0$ has at least one zero in **C***, if $n \geq 1$.*

Hint : Assume not. Choose R so large that

$$|p(z) - z^n| < |z^n| \text{ for all } |z| = R$$

consider the closed curve

$$\gamma(t) = R \exp(2\pi i t) \text{ for } t \in [0, 1]$$

and combine Theorem 17(b) with Exercise 17.

19. Show that the complex valued function

$$f(z) = z^{17} - 16i \sin\left(93|z|^{22}\right) z^{12} + 2$$

has a zero in the disc B(0,2) . [For heaven's sake, don't try to compute it!]

20. Let A be an invertible 3 x 3 matrix whose entries all are ≥ 0. Show that A has a positive eigenvalue and a corresponding eigenvector (v_1, v_2, v_3) such that $v_1 \geq 0$, $v_2 \geq 0$, $v_3 \geq 0$.

This result is known as the *Perron-Frobenius Theorem*, and it is true for $n \times n$ matrices.

Hint: Consider the compact convex set

$$\left\{(x_1, x_2, x_3) \in \mathbf{R}^3 \mid x_1 + x_2 + x_3 = 1 , \ x_1 \geq 0 , \ x_2 \geq 0 , \ x_3 \geq 0\right\}$$

and the mapping

$$x \rightarrow \frac{Ax}{(Ax)_1 + (Ax)_2 + (Ax)_3}$$

21. Let $\sqrt{\ } : \mathbf{C} \backslash\ ] - \infty, 0] \rightarrow \mathbf{C}$ be the principal square root of z. For which values of z does the equation $\sqrt{z^2} = z$ hold?

22. Consider the square root $\sqrt{z^2 - 1}$ from Example 22(c). Find $\sqrt{z^2 - 1}$ for z on the imaginary axis.

Show that $\sqrt{z^2 - 1}$ has a limit as $z \rightarrow 0$ through the open upper half plane, and find it. Same question when $z \rightarrow 0$ through the open lower half plane.

23. There are many ingenious ways of showing the non-existence of a continuous $\sqrt{z}$ on the unit circle S^1. For the benefit of the reader we will below sketch 4 different methods. They all proceed by contradiction. So we assume that there exists a branch r of $\sqrt{z}$ on S^1, i.e. a continuous function r satisfying $r(z)^2 = z$ for all $z \in S^1$. We may without loss of generality assume that $r(1) = 1$.

Method 1: The one from Example 22(b).

Method 2: Show that the function $\phi(z) := r(z)/r(-z)$ is a constant (indeed $\phi(z)^2 = -1$). Derive a contradiction.

Method 3: Consider the smooth function $f(t) := r(\exp(it))$ for $t \in \mathbf{R}$. Differentiate the identity $f(t)^2 = \exp(it)$).

Method 4: (α) Show first that r is a continuous injective mapping.

(β) Show that $r(S^1) = S^1$ [Either by a point set topology argument ($r(S^1)$ is homeomorphic to S^1 and hence not starshaped (Cf Exercise 4) , or by showing that $r(S^1)$ is a subgroup of the multiplicative group S^1].

(γ) Choose a point $z_0 \in S^1$ such that $r(z_0) = -1$. Obtain a contradiction using the injectivity of r^2.

24. Does there exist a branch of the nth root of z on S^1 when $n > 0$? Exhibit a specimen if your answer is yes.

25. Let $h : S^1 \to \mathbf{C}\backslash\{0\}$ be continuous, and consider the closed curve $\gamma(t) := h(\exp(it))$ for $t \in [0, 2\pi]$.

(α) Show that h can be written in the form $h(z) = z^n \exp(\phi(z))$, where n is an integer and $\phi \in C(S^1)$. Hint: Let F be a continuous logarithm of $t \to h(e^{it})$ and $n = Ind_\gamma(0)$, so that

$$h(e^{it}) = e^{F(t)} = e^{int} e^{F(t) - int}$$

Note that the exponent $\psi(t) := F(t) - int$ satisfies that $\psi(2\pi) = \psi(0)$, and so defines a function ϕ on S^1.

(β) Show that $Ind_\gamma(0)$ is even if h is even.

(γ) Show that $Ind_\gamma(0)$ is odd if h is odd. In particular, if h is odd then $Ind_\gamma(0) \neq 0$.

26. We will in this exercise present a proof of the 2-dimensional version of the so-called

Brouwer's Theorem on Invariance of Domain:

Let Ω be an open subset of the complex plane, and let $f : \Omega \to \mathbf{C}$ be continuous and 1-1. Then $f(\Omega)$ is open in $\mathbf{C}$.

(α) Verify that the theorem is a consequence of the following

Lemma. If the function $f : B[0, 1] \to \mathbf{C}$ is continuous and 1-1, then $f(0)$ is an interior point of $f(B[0, 1])$.

We proceed with a proof of the lemma:

(β) Show that we may assume that $f(0) = 0$, and that then $r := dist(0, f(S^1)) > 0$.

(γ) Show that it suffices to lead the following assumption to a contradiction: There exists a point $z_0 \in B(0, r)\backslash f(B[0, 1])$.

(δ) Consider now the closed curve $\gamma(t) := f(e^{it})$ for $t \in [0, 2\pi]$. Show that $0 = Ind_\gamma(z_0) = Ind_\gamma(0)$.

(ϵ) Show that $F : S^1 \times [0, 1] \to \mathbf{C}\backslash\{0\}$, given by

$$F(z, s) := f\left(\frac{z}{1+s}\right) - f\left(\frac{-sz}{1+s}\right)$$

is a homotopy in $\mathbf{C}\backslash\{0\}$ between $f|_{S^1}$ and the odd function

$$h(z) := F(z, 1) = f\left(\frac{z}{2}\right) - f\left(\frac{-z}{2}\right)$$

(ζ) Use Exercise 25(γ) to arrive at the desired conclusion.

27. (α) Show that there at any given time are opposite ends of the earth which have the same weather, i.e. the same temperature and the same barometer pressure. This is a meteorological interpretation of the 2-dimensional version of

Borsuk-Ulam's Theorem.

Let $S^2 := \{x \in \mathbf{R}^3 \mid |x| = 1\}$ be the unit sphere. If $f : S^2 \to \mathbf{C}$ is continuous then there exists an $x_0 \in S^2$ such that $f(x_0) = f(-x_0)$.

(β) Prove the Borsuk-Ulam theorem. Hint : Consider the map $h : B[0,1] \to \mathbf{C}$ given by

$$h(x + iy) := f\left(x, y, \sqrt{1 - x^2 - y^2}\right) - f\left(-x, -y, -\sqrt{1 - x^2 - y^2}\right)$$

and the closed curve $\gamma(t) := h(\exp{(it)})$ in $\mathbf{C}$.

(γ) Prove the following : If a balloon is deflated and laid on the floor, then there exist two antipodal points which end up over the same point of the floor.

28. (Cf Proposition 20) Let $f \in Hol(\Omega)$ and assume that the zeros of f all are isolated. Let r be a continuous square root of f.

Show that $r \in Hol(\Omega)$.

Chapter 4 Basic Theory of Holomorphic Functions

While Chapter III essentially treats continuous functions and a bit of topology of the complex plane, this chapter specializes to holomorphic functions and their surprising properties, derived from the Cauchy integral theorem. The contents of this chapter are dealt with in all introductory text books on complex function theory.

Section 1 The Cauchy-Goursat integral theorem

The main results of this paragraph are the Cauchy-Goursat integral theorem and some of its surprising consequences: The integral of a holomorphic function f around a closed curve, the interior of which is contained in the domain of definition of f, is zero. A holomorphic function is analytic, i.e. it can be expanded in a power series.

Theorem 1 (Cauchy-Goursat).

Let f be holomorphic in an open subset Ω of the complex plane. Let $a, b, c \in \Omega$ and assume that the triangle $\triangle$ with vertices a, b and c is contained in Ω. Then

$$\int_{\partial\triangle} f(z)dz = 0$$

Proof:

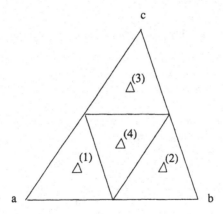

Divide the triangle $\triangle$ as in the figure, using the vertices and the midpoints of the segments $[a, b]$, $[b, c]$ and $[c, a]$. We get 4 smaller triangles $\triangle^{(1)}, \triangle^{(2)}, \triangle^{(3)}$ and $\triangle^{(4)}$ as shown in the figure. Orienting all the triangles in the counter clockwise direction it is a matter of direct verification that

$$\int_{\partial \Delta} f(z)dz = \sum_{i=1}^{4} \int_{\partial \Delta^{(i)}} f(z)dz$$

Denoting the absolute value of the left hand side for α, we have for at least one small triangle $\Delta^{(i)}$, call it Δ_1, that

$$\left| \int_{\partial \Delta_1} f dz \right| \geq \frac{\alpha}{4}$$

Note that $l(\partial \Delta_1) = \frac{1}{2}l(\partial \Delta)$ where l denotes length of the path in question.

We can repeat the process to successively obtain triangles $\Delta_n \subseteq \Delta_{n-1}$ such that $l(\partial \Delta_n) = 2^{-n}l(\partial \Delta)$ and

$$\left| \int_{\partial \Delta_n} f dz \right| \geq 4^{-n} \alpha$$

There is a point, say z_0, in the intersection $\bigcap_{n=1}^{\infty} \Delta_n$. f is differentiable at z_0, so given $\beta > 0$ we have from a certain n on, in the entire triangle Δ_n, that

$$(*) \quad |f(z) - f(z_0) - f'(z_0)(z - z_0)| \leq \beta |z - z_0|$$

The constant function $f(z_0)$ and the function $z - z_0$ have primitives, so (by Proposition II.10.(e)) the line integrals of these along any closed path vanish. Thus for any Δ_n where (*) holds:

$$4^{-n} \alpha \leq \left| \int_{\Delta_n} f dz \right| \leq \beta(l(\partial \Delta_n))^2 = \beta 4^{-n}(l(\partial \Delta))^2$$

where we have used that $z_0 \in \Delta_n$ and that $|z - z_0| \leq l(\partial \Delta_n)$ for any $z \in \partial \Delta_n$. Now, $\beta(l(\partial \Delta))^2 \geq \alpha$. Since this holds for any $\beta > 0$ we conclude that $\alpha = 0$. $\qquad \square$

By a bit of ingenuity we can get a stronger looking version of Theorem 1 out by allowing the function f to have singularities :

Corollary 2.

Let Ω be an open subset of $\mathbf{C}$, and let f be holomorphic in $\Omega \backslash \{z_0\}$. Assume furthermore that $(z - z_0)f(z) \to 0$ as $z \to z_0$. Then

$$\int_{\partial \Delta} f(z)dz = 0$$

for any triangle $\triangle$ in Ω such that $z_0 \in \Omega \backslash \partial \triangle$.

Proof:

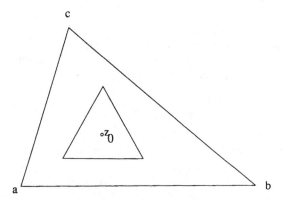

Let δ be a small triangle inside of $\triangle$, containing z_0 in its interior. Subdividing $\triangle \backslash \delta$ into smaller triangles we get by the Cauchy-Goursat theorem that

$$\int_{\partial \triangle} f(z)dz = \int_{\partial \delta} f(z)dz$$

Let us now consider a sequence $\delta_1, \delta_2, \cdots$ of equilateral triangles, each with z_0 as its center, and with diameters shrinking to 0. By elementary geometry, when we choose δ_n so that $dist(z_0, \partial \delta_n) = 1/n$ then $l(\partial \delta_n) = 6\sqrt{3}/n$. By assumption

$$|f(z)| \leq \frac{c(z)}{|z - z_0|} \quad \text{where} \quad c(z) \to 0 \quad \text{as} \quad z \to z_0$$

Let $\epsilon > 0$ be given. Choosing n so large that $|c(z)| \leq \epsilon$ for all $z \in \partial \delta_n$ we get

$$\left| \int_{\partial \triangle} f \right| = \left| \int_{\partial \delta_n} f \right| \leq \sup_{\partial \delta_n} \{|f|\} l(\partial \delta_n) \leq \frac{\epsilon}{(1/n)} \frac{6\sqrt{3}}{n} = 6\sqrt{3}\epsilon$$

The statement of the corollary follows since this is true for any $\epsilon > 0$. $\square$

From Corollary 2 we deduce a remarkable and important regularity property: A holomorphic function cannot possess a minor singularity at a point. If it has a point singularity it is a serious one. This result is called "The removable singularity theorem" or "Riemann's extension theorem":

Theorem 3 (The removable singularity theorem).

Let z_0 be a point in an open subset Ω of the complex plane, and let $f : \Omega \backslash \{z_0\} \to \mathbf{C}$ be holomorphic in $\Omega \backslash \{z_0\}$.

If $(z - z_0)f(z) \to 0$ as $z \to z_0$ [in particular if f is bounded near z_0], then f can be extended to a function which is holomorphic in all of Ω.

Proof:

Without loss of generality we take $z_0 = 0$ in the proof. Let $\triangle$ be a triangle, contained in Ω, with $z_0 = 0$ in its interior, and its boundary $\partial \triangle$ positively oriented (Cf Example III.13) so that (Theorem III.17(d))

$$1 = \frac{1}{2\pi i} \int\limits_{\partial \triangle} \frac{dw}{w - z} \text{ for any } z \in \text{int}(\triangle)$$

For any fixed $z \in \text{int}(\triangle) \backslash \{0\}$ each of the two singularities 0 and z of the function

$$w \to \frac{f(w) - f(z)}{w - z} \text{ on } \Omega \backslash \{0, z\}$$

satisfies the condition of Corollary 2. Subdividing $\triangle$ into two triangles, each with only one of the singularities, we get

$$\int\limits_{\partial \triangle} \frac{f(w) - f(z)}{w - z} dw = 0$$

which implies that

$$f(z) = \frac{1}{2\pi i} \int\limits_{\partial \triangle} \frac{f(w)}{w - z} dw \text{ for any } z \in \text{int}(\triangle) \backslash \{0\}$$

The right side is analytic and hence holomorphic off $\partial \triangle$ (Lemma II.13), so the desired extension F may unambiguously be defined by

$$F(z) := \begin{cases} f(z) & \text{for } z \in \Omega \backslash \{0\} \\ \frac{1}{2\pi i} \int\limits_{\partial \triangle} \frac{f(w)}{w-z} dw & \text{for } z \in \text{int}(\triangle) \end{cases}$$

$\square$

We see that Corollary 2 actually does not strengthen Theorem 1, because the exceptional point is not a singularity after all! The corresponding result is not true in one real variable: The function $t \to |t|$ is not differentiable at 0.

During the proof of Theorem 3 we noticed the remarkable fact that a holomorphic function is analytic. We shall in Theorem 8 below prove that its power series converges in the largest possible disc.

As a by-product of "The removable singularity theorem" we note the following example which will be used several times, both in the near and more distant future.

Example 4.

If f is holomorphic in the open set Ω and $a \in \Omega$ then the function

$$g(z) := \begin{cases} \frac{f(z)-f(a)}{z-a} & \text{for } z \in \Omega \backslash \{a\} \\ f'(a) & \text{for } z = a \end{cases}$$

is holomorphic in all of Ω.

Theorem 5 (The Cauchy integral theorem).

Let f be holomorphic in an open starshaped subset Ω of the complex plane. Then $\int_\gamma f(z)dz = 0$ for each closed path γ in Ω.

Proof: When we combine the Cauchy-Goursat theorem with Lemma II.12 we see that f has a primitive in Ω, so we may refer to Proposition II.10(e). □

Theorem 6 (The Cauchy integral formula).

Let f be holomorphic in an open starshaped set Ω of the complex plane. Then we have for any closed path γ in Ω the formula

$$Ind_\gamma(z)f(z) = \frac{1}{2\pi i} \int_\gamma \frac{f(w)}{w-z}dw \text{ for all } z \in \Omega \backslash \gamma^*$$

Proof:
Fix $z \in \Omega \backslash \gamma^*$. Then the function

$$w \to \begin{cases} \frac{f(w)-f(z)}{w-z} & \text{for } w \in \Omega \backslash \{z\} \\ f'(z) & \text{for } w = z \end{cases}$$

is according to Example 4 holomorphic in Ω, so by the Cauchy integral theorem its integral along γ is 0, i.e.

$$\int_\gamma \frac{f(w)-f(z)}{w-z}dw = 0$$

Recalling the formula (Theorem III.17(d))

$$Ind_\gamma(z) = \frac{1}{2\pi i} \int_\gamma \frac{dw}{w-z}$$

we get the theorem. □

47

As a special case we note *the Cauchy integral formula for a disc:*

Theorem 7.
If f is holomorphic in a neighborhood of the closed disc $B[z_0, r]$, then we have for each $z \in B(z_0, r)$ the formula

$$f(z) = \frac{1}{2\pi i} \int\limits_{|w-z_0|=r} \frac{f(w)}{w - z} dw$$

The purpose of the next chapter is to extend the Cauchy integral theorem and the Cauchy integral formula to more general domains than starshaped. The following remark (which will not be used later) points in another direction.

Remark.
There is an extension of the Cauchy integral formula to general functions. It is often called the *Cauchy-Green formula* since Green's theorem is used to prove it (For a proof, see e.g. [Hö;Theorem 1.2.1]):

Let ω be a bounded open domain in $\mathbf{C}$ with a smooth positively oriented boundary γ. If u is $\mathbf{C}^1$ in a neighborhood of $\bar{\omega}$ then

$$u(z) = \frac{1}{2\pi i} \int\limits_\gamma \frac{u(w)}{w - z} dw - \frac{1}{\pi} \int\limits_\omega \frac{\frac{\partial u}{\partial \bar{z}}(x)}{x - z} dm(x) \text{ for } z \in \omega$$

where dm denotes Lebesgue measure on $\mathbf{C} = \mathbf{R}^2$.

Note that the last term on the right hand side drops out if u is holomorphic, so that the formula in that case reduces to the ordinary Cauchy integral formula.

A version of the Cauchy integral theorem can be derived from the Cauchy-Green theorem (replace u by $(w - z)u(w)$ etc.):

$$\frac{1}{2\pi i} \int\limits_\gamma u(z) dz = \frac{1}{\pi} \int\limits_\omega \frac{\partial u}{\partial \bar{z}}(x) dm(x)$$

There is a striking contrast between $\mathbf{R}$- and $\mathbf{C}$-differentiability: There are examples of functions on $\mathbf{R}$ with the property that f is continuously differentiable and f' is not differentiable anywhere. But as we noticed after Theorem 3 any $\mathbf{C}$-differentiable function, i.e. a holomorphic function, is in fact even analytic and hence in particular infinitely often differentiable. In the future we will use the words analytic and holomorphic interchangeably. What is new in Theorem 8 below is that the power series converges in the largest possible disc.

Theorem 8.

Let f be holomorphic in an open subset Ω of the complex plane. Then f is analytic in Ω; furthermore, for any $z_0 \in \Omega$ the power series expansion of f around z_0 converges in the largest open disc in Ω around z_0.

In particular, f is infinitely often $\mathbf{C}$-differentiable and all derivatives $f', f'', \cdots$ are holomorphic.

Also $f \in C^\infty(\Omega)$, i.e. partial derivatives of all orders with respect to the real variables x and y exist and are continuous.

Proof: Combine Lemma II.13 and Theorem 7 with the uniqueness of power series expansions. For the last statement we note that

$$\frac{\partial f}{\partial x} = f' \text{ and } \frac{\partial f}{\partial y} = if'$$

from which we by induction get

$$\frac{\partial^{n+m} f}{\partial x^m \partial y^n} = i^n f^{(n+m)}$$

$\square$

It may be remarked that Theorem 8 does not hold for an analytic function of a real variable. To take an example, the function

$$f(x) := \frac{1}{1+x^2} \text{ for } x \in \mathbf{R}$$

is analytic on all of the real line, but its power series expansion

$$f(x) = 1 - x^2 + x^4 - x^6 + \cdots$$

converges only for $-1 < x < 1$. The explanation is that if we view f as a function of a complex variable

$$f(z) := \frac{1}{1+z^2} \text{ for } z \in \mathbf{C}$$

then f is not holomorphic = analytic on all of the complex plane, but only on the domain $\mathbf{C}\backslash\{i, -i\}$, and the biggest disc around 0 in that domain is $B(0,1)$.

We next present a converse to the Cauchy-Goursat theorem, viz. Morera's theorem, which is often expedient in determining whether a given function is holomorphic. It can be used in situations where a direct resort to the definition - estimation of difference quotients etc. - is hopeless or at least very complicated. For a more thorough discussion of Morera's theorem we refer to the beautiful article [Za]. For the latest news see [Gl].

Theorem 9 (Morera's theorem).

Let $f : \Omega \to \mathbf{C}$ be a continuous function on an open subset Ω of the complex plane. Assume that each point $z \in \Omega$ has a neighborhood $U_z \subseteq \Omega$ with the property that $\int_\Delta f(z)dz = 0$ for all triangles Δ contained in U_z.

Then f is holomorphic in Ω.

Proof: Let $z \in \Omega$ and choose $r > 0$ so small that $B(z,r) \subseteq U_z$. It suffices to prove that f is holomorphic in $B(z,r)$ for each $z \in \Omega$. Now f has by Lemma II.12 a primitive F in $B(z,r)$. Being $\mathbf{C}$-differentiable means that F is holomorphic. By Theorem 8 so is its derivative f. $\qquad\square$

Section 2 Selected consequences of the Cauchy integral formula

It is a sad fact from the theory of functions of a real variable that a uniform limit of a sequence of differentiable functions need not be differentiable, although it is continuous. Indeed, there exists a continuous nowhere differentiable function on [0,1], and (by the classical Weierstrass approximation theorem) it is a uniform limit of a sequence of polynomials.

The situation is quite the contrary for holomorphic functions, as demonstrated by the following surprising result, in which no assumptions are made on the derivatives:

Theorem 10 (Weierstrass' theorem).

Let $f_1, f_2, \cdots$ be a sequence of functions which are holomorphic in an open subset Ω of the complex plane. Assume that the sequence converges, uniformly on each compact subset of Ω, to a function f.

Then f is also holomorphic in Ω. Furthermore, the sequence $\{f_n'\}$ of derivatives converges, again uniformly on each compact subset of Ω, to f'.

We shall later (Exercise VIII.10) meet a stronger version of Weierstrass' theorem, viz. Vitali-Porter's theorem. The assumption about uniform convergence is important; pointwise convergence does not suffice. For a discussion see [Dn] and [Za;§11].

Proof:

Let $z_0 \in \Omega$ and choose $r > 0$ be so small that $B[z_0,r] \subseteq \Omega$. According to the Cauchy integral formula (Theorem 7) we have

$$(*) \quad f_n(z) = \frac{1}{2\pi i} \int_{|w-z_0|=r} \frac{f_n(w)}{w - z} dw \quad \text{for all } z \in B(z_0,r)$$

and going to the limit $n \to \infty$ we get

$$(**) \quad f(z) = \frac{1}{2\pi i} \int\limits_{|w-z_0|=r} \frac{f(w)}{w-z} dw \quad \text{for all} \quad z \in B(z_0, r)$$

which [by Lemma II.13] shows that f is analytic and hence holomorphic in $B(z_0, r)$. Now $f \in Hol(\Omega)$, the disc being arbitrary.

Concerning the last statement we may now assume that $f = 0$ [if necessary we replace f_n by $f_n - f$] .

By a compactness argument it suffices to show that $\{f'_n\}$ converges uniformly to 0 as $n \to \infty$ on any ball $B\left(z_0, \frac{r}{2}\right)$ such that $B\left[z_0, \frac{r}{2}\right] \subseteq \Omega$. Differentiating (*) we get [directly or using Lemma II.13] that

$$f'_n(z) = \frac{1}{2\pi i} \int\limits_{|w-z_0|=r} \frac{f_n(w)}{(w-z)^2} dw \quad \text{for all} \quad z \in B\left(z_0, \frac{r}{2}\right)$$

which we estimate as follows for $z \in B\left(z_0, \frac{r}{2}\right)$ [Cf Proposition II.10.(a)]:

$$\left| f'_n(z) \right| \le \frac{1}{2\pi} \frac{2\pi r}{(r/2)^2} \sup_{|w-z_0|=r} \{|f_n(w)|\} = \frac{4}{r} \sup_{|w-z_0|=r} \{|f_n(w)|\}$$

so

$$\sup_{B(z_0, r/2)} \{|f'_n|\} \le \frac{4}{r} \sup_{|w-z_0|=r} \{|f_n(w)|\}$$

The circle $\{w \in \mathbf{C} \mid |w - z_0| = r\}$ is a compact subset of Ω, so the right hand side converges to 0 as $n \to \infty$. $\qquad \Box$

An easy corollary is the fact (Proposition II.4) that a power series in its disc of convergence defines a holomorphic function and that its derivative can be found by term by term differentiation (See Exercise 1). More generally one can apply Weierstrass' theorem to infinite series, the terms of which are holomorphic functions. Example: The Riemann ζ-function (Exercise 26 below and Chapter XI.§1).

We next generalize Proposition 1.6.

Theorem 11 (The unique continuation theorem).

Let Ω be an open connected subset of the complex plane, and let $f, g \in Hol(\Omega)$.

(α) If the set of points $z \in \Omega$ where $f(z) = g(z)$ has a limit point in Ω, then $f = g$.

(β) In particular, the zeros of f have no limit points in Ω, unless f is identically 0.

We remind the reader, that a point p is a *limit point* of a set A, if every neighborhood of p contains at least one point of A distinct from p.

We emphasize that the limit point from (α) and (β) must belong to Ω; limit points outside of Ω do not suffice.

Proof:

(β) is a consequence of (α), so it suffices to prove (α), and it is even enough to do so with $g = 0$. Consider the set

$$N := \left\{ a \in \Omega | f^{(n)}(a) = 0 \text{ for all } n = 0, 1, 2, \cdots \right\}$$

Its complement in Ω , i.e.

$$\Omega \backslash N = \left\{ z \in \Omega | \text{There exists an n such that } f^{(n)}(z) \neq 0 \right\}$$

is open because each $f^{(n)}$ is continuous. On the other hand, if $a \in N$, then we see from the power series expansion

$$f(z) = \sum_{n=0}^{\infty} \frac{f^{(n)}(a)}{n!} (z - a)^n$$

that f (and hence each of its derivatives, too) is identically 0 in a neighborhood of a, so N is also open. By connectedness of Ω either $N = \Omega$ or $N = \emptyset$. It suffices to prove that the limit point $z_0 \in \Omega$ belongs to N, so that the first possibility takes place. We shall in other words show that in the power series expansion

$$f(z) = \sum_{k=0}^{\infty} a_k (z - z_0)^k$$

of f around z_0 all the coefficients $a_0, a_1, a_2, \cdots$ are 0.

But that is part of Proposition I.6. $\square$

The unique continuation theorem is amazing : It tells us that a holomorphic function is determined by its values on e.g. any curve segment, be it ever so small; and that we cannot change the values of the function in one part of the domain without it being felt everywhere.

As an illustrative example we mention that Log thus can be characterized uniquely as the holomorphic function on $C \backslash] - \infty, 0]$ that extends the ordinary real logarithm on $]0, \infty[$.

The unique continuation theorem implies that holomorphic functions obey the so-called "principle of permanence of functional relationships", i.e. if holomorphic functions in a part of a region satisfy a certain relationship, then they do so everywhere. As an example we infer from the well known relation $\sin^2 x + \cos^2 x = 1$, which is true for all real x, that the same relation holds everywhere, i.e. $\sin^2 z + \cos^2 z = 1$ for all complex z (Exercise 16).

Definition 12.

Let f be holomorphic in open subset Ω of the complex plane and let $z_0 \in \Omega$. Then f can in exactly one way be expanded in a power series (in the largest open ball B in Ω) around z_0:

$$f(z) = \sum_{k=0}^{\infty} a_k(z - z_0)^k \ \text{for} \ z \in B$$

Clearly f has a zero at z_0 if and only if $a_0 = 0$. We define the order *of the zero as*

$$\sup \{n + 1 \,|\, a_0 = a_1 = \cdots = a_n = 0\} \in \mathbb{N} \cup \{\infty\}$$

An equivalent characterization that does not involve the power series expansion, is that the order is the biggest n such that

$$f(z_0) = f'(z_0) = \cdots = f^{(n)}(z_0) = 0$$

We leave it to the reader to prove, that if f has a zero of order $n < \infty$, then f can be factorized as

$$f(z) = (z - z_0)^n g(z) \ \text{where} \ g \in Hol(\Omega) \ \text{and} \ g(z_0) \neq 0$$

If f has a zero of infinite order then f vanishes identically near the zero (and hence on all of the connected component of Ω that contains z_0).

Section 3 The open mapping theorem

The following continuous version of Rouché's theorem will be used in the proofs of the Open Mapping Theorem and of Bloch-Landau's theorem (Theorem VII.6). Rouché's theorem will be discussed in more details in Chapter IX.

Theorem 13 (Rouché's theorem).

Let $F \in C(B[0,r])$ and let H be holomorphic on a neighborhood of $B = B[0,r]$. Assume that

$$(*) \quad |F(\zeta) - H(\zeta)| \leq |H(\zeta)| \ \text{for all} \ \zeta \in \partial B$$

and that H has a zero somewhere in $B[0,r]$.

Then F, too, has a zero in $B[0,r]$. If the inequality (*) is strict then F has a zero in the open ball $B(0,r)$.

Proof: (The proof is adapted from [Ts]). We shall derive a contradiction from the assumption that F never vanishes in $B[0,r]$. Let

$$\gamma(t) := re^{it} \ \text{for} \ 0 \leq t \leq 2\pi$$

Since F never vanishes we may divide (*) by F to get that

$$\left|1 - \frac{H(\zeta)}{F(\zeta)}\right| \leq \left|\frac{H(\zeta)}{F(\zeta)} - 0\right| \quad \text{for } \zeta \in \partial B$$

which shows that $H(\zeta)/F(\zeta)$ is closer to 1 than to 0. Hence $\Re\{H(\zeta)/F(\zeta)\} \geq \frac{1}{2}$, so we may form $Log\{H(\zeta)/F(\zeta)\}$ for $|\zeta| = r$. Letting ϕ be a continuous logarithm of $F \circ \gamma$ we get that $\phi + Log\{H \circ \gamma/F \circ \gamma\}$ is a continuous logarithm of $H \circ \gamma$, and so $Ind_{H \circ \gamma}(0) = Ind_{F \circ \gamma}(0)$.

The map Γ, defined by

$$\Gamma(t,s) := F\big(sre^{it}\big) \quad \text{for } (t,s) \in [0, 2\pi] \times [0,1]$$

is a homotopy in $\mathbf{C}\backslash\{0\}$ between the constant map $F(0)$ and $F \circ \gamma$, so

$$Ind_{H \circ \gamma}(0) = Ind_{F \circ \gamma}(0) = Ind_{F(0)}(0) = 0$$

We arrive at the desired contradiction by proving that $Ind_{H \circ \gamma}(0) \neq 0$:

We may write H in the form

$$H(z) = (z - z_1)(z - z_2) \cdots (z - z_n)g(z)$$

where $z_1, z_2, \cdots, z_n$ are the zeros of H in $B(0,r)$ and where g is holomorphic in a neighborhood of $B[0,r]$ and never vanishes in $B[0,r]$. In particular

$$\frac{H'(z)}{H(z)} = \sum_{j=1}^{n} \frac{1}{z - z_j} + \frac{g'(z)}{g(z)}$$

Now,

$$Ind_{H \circ \gamma}(0) = \frac{1}{2\pi i} \int\limits_{H \circ \gamma} \frac{dz}{z} = \frac{1}{2\pi i} \int\limits_{\gamma} \frac{H'(z)}{H(z)} dz$$

$$= \sum_{j=1}^{n} \frac{1}{2\pi i} \int\limits_{\gamma} \frac{dz}{z - z_j} + \frac{1}{2\pi i} \int\limits_{\gamma} \frac{g'(z)}{g(z)} dz = n + 0 = n > 0$$

$\square$

Theorem 14 (The open mapping theorem).

Let f be a non-constant holomorphic function on an open connected subset Ω of the complex plane. Then $f(\Omega)$ is open.

Proof: It suffices to prove that there for each $z_0 \in \Omega$ exists a disc, contained in $f(\Omega)$, around $f(z_0)$.

Consider the function $H(z) := f(z) - f(z_0)$ for $z \in \Omega$. By the Unique Continuation Theorem 11(β) there is a ball $B[z_0, r]$ in which the only zero of H is $z = z_0$. In particular

$$R := \inf_{|\zeta - z_0| = r} \{|H(\zeta)|\} > 0$$

Let now w be any point such that $|w - f(z_0)| < R$, and let us apply Rouché to the two functions $F(z) = f(z) - w$ and H : For $|\zeta - z_0| = r$ we get

$$|F(\zeta) - H(\zeta)| = |f(z_0) - w| < R \leq |H(\zeta)|$$

so by Rouché F has a zero in $B[z_0, r]$, i.e. w belongs to the range of f. Since w was arbitrary in $B(f(z_0), R)$, we have proved that any element of $B(f(z_0), R)$ belongs to the range of f, or in other words that $B(f(z_0), R) \subseteq f(\Omega)$. □

A topological proof of Theorem 14 can be found in [CS].

An immediate consequence of The Open Mapping Theorem is

Corollary 15 (The Maximum Modulus Principle).

Let f be holomorphic in open connected subset Ω of $\mathbf{C}$ and assume that $|f|$ has a local maximum in Ω. Then f is constant in Ω.

Proof: Make a sketch! □

In applications the following consequence is useful.

Corollary 16 (The Weak Maximum Principle).

Let Ω be a bounded, open subset of $\mathbf{C}$ and let f be holomorphic on Ω and continuous on the closure of Ω. Then $|f|$ takes its supremum value at a boundary point of Ω.

Proof: Since $f \in C(\overline{\Omega})$ and $\overline{\Omega}$ is compact, $|f|$ takes its supremum value M in $\overline{\Omega} = \Omega \cup \partial\Omega$. Let us assume that it happens at an interior point $z_0 \in \Omega$ and then deduce that $|f|$ takes the same value at some boundary point, too.

By the Maximum Modulus Principle f is constant in the biggest open ball $B(z_0, R)$ in Ω with center z_0. Since the ball is the biggest, its boundary contains at least one point $z\prime$ of $\partial\Omega$. By continuity $|f(z')| = M$. □

We shall in the chapters XII and XIII extend these maximum principles to harmonic and subharmonic functions.

As an example of how to use the maximum principles we shall prove Schwarz' lemma.

Theorem 17 (Schwarz' lemma).

Let $f : B(0,1) \to B[0,1]$ be holomorphic with $f(0) = 0$. Then

(α) $|f(z)| \le |z|$ for all $z \in B(0,1)$, and $|f'(0)| \le 1$.

(β) If there exists a $z_0 \in B(0,1)\backslash\{0\}$ such that $|f(z_0)| = |z_0|$, or if $|f'(0)| = 1$, then f has the form $f(z) = cz$ for all $z \in B(0,1)$, where c is a constant of absolute value 1.

Proof: The function

$$g(z) := \begin{cases} \frac{f(z)}{z} & \text{for } z \in B(0,1)\backslash\{0\} \\ f'(0) & \text{for } z = 0 \end{cases}$$

is holomorphic in $B(0,1)$ (Cf. Example 4 above). For $|z| < r < 1$ we infer from the Weak Maximum Principle that

$$|g(z)| \le \max_{|w|=r} \{|g(w)|\} = \max_{|w|=r} \left\{ \frac{|f(w)|}{r} \right\} \le \frac{1}{r}$$

and letting $r \to 1_-$ we get $|g(z)| \le 1$, i.e. (α).

In each of the cases of (β) the function $|g|$ assumes its supremum in $B(0,1)$, so by The Maximum Modulus Theorem g is a constant function, $g(z) = c$. And $|c| = |g(z_0)| = 1$. This proves (β). $\qquad\square$

The conclusion $|f(z)| \le |z|$ of Schwarz' lemma expresses that f does not increase modulus. It implies in particular that if z ranges over the subdisc $B(0,r)$ then the values $f(z)$ also lie in that disc. Such considerations form the basis for the so-called *Principle of Subordination*. For more information on that topic we refer the reader to the paper [Mac].

Another important application of Schwarz' lemma is to the uniqueness question of conformal mappings, so we will return to Schwarz' lemma during our treatment of the Riemann Mapping Theorem.

We have already earlier (Theorem III.5) observed that any continuous function g satisfying $(\exp \circ g)(z) = z$, i.e. any continuous logarithm, automatically is holomorphic. And we have claimed that an assumption in Proposition III.20 (that f never vanishes) is superfluous. Both these results are special cases of our next theorem.

Theorem 18.

Let $g : \Omega \to \mathbb{C}$ be a continuous function on an open subset Ω of the complex plane, and let f be a non-constant holomorphic function, defined on an open connected subset of the complex plane containing $g(\Omega)$.

If the composite map $f \circ g$ is holomorphic on Ω, then so is g.

Proof: Since holomorphy is a local property we may in the proof assume that Ω is connected. Let $a \in \Omega$ be arbitrary. We shall show that g is complex differentiable at a.

Since f is holomorphic around $g(a)$ we can write

$$f(w) - f(g(a)) = (w - g(a))^k H(w)$$

where H is holomorphic and $H(g(a)) \neq 0$. So there is an open disc D around $g(a)$ in which H never vanishes. By Proposition III.21 H has a holomorphic k^{th} root f_0 on D. So we write

$$f(w) - f(g(a)) = [(w - g(a))f_0(w)]^k \quad \text{for all } w \text{ near } g(a)$$

Here k is an integer ≥ 1, $f_0 \in Hol(B(g(a), \delta))$ for some $\delta > 0$ and $f_0(g(a)) \neq 0$.

Replacing w by $g(z)$ and using the abbreviation h for the holomorphic function $h = f \circ g - f(g(a))$, we get for z in a neighborhood of a, say for z in $B(a, \epsilon) \subseteq \Omega$, that

$$h(z) = [(g(z) - g(a))f_0(g(z))]^k$$

so we see that the bracket on the right hand side is a continuous k^{th} root r of the holomorphic function h on the left hand side. There are now two cases:

1) h is identically 0 in $B(a, \epsilon)$. In this case $g(z) = g(a)$ near $z = a$, so g is clearly complex differentiable at a.

2) h is not identically 0 in $B(a, \epsilon)$. In this case we know by the Unique Continuation Theorem (Theorem 11) that a is an isolated zero for h. According to Proposition III.20 the root r is holomorphic in $B(a, \epsilon) \backslash h^{-1}(0)$, so a is an isolated singularity of r. But r is continuous in $B(a, \epsilon)$, so by the Removable Singularity Theorem (Theorem 3) r is actually holomorphic in a neighborhood of a.

Dividing the identity $r(z) = (g(z) - g(a))f_0(g(z))$ by $z - a$ and noting that $r(a) = 0$ we get

$$\frac{r(z) - r(a)}{z - a} = \frac{g(z) - g(a)}{z - a} f_0(g(z))$$

The left hand side has a limit as $z \to a$, and, since $f_0(g(a)) \neq 0$, so does the difference quotient of g, i.e. g is differentiable at a. $\quad\square$

Terminology 19.

A function is said to be univalent, *if it is holomorphic and injective.*

Theorem 20.

If f is univalent in an open set Ω, then $f(\Omega)$ is open and f^{-1} is holomorphic on $f(\Omega)$.

Proof: Being open, Ω is a union of open balls. f is injective so it is not constant on any such ball. Hence $f(\Omega)$ is open by the Open Mapping Theorem (Theorem 14). Since f is an open map, f^{-1} is continuous, so we can apply Theorem 18 with $g = f^{-1}$. $\quad\square$

The reader might note that a result corresponding to Theorem 20 is false for functions of one real variable: The inverse of a differentiable function need not be differentiable (Example $f(x) = x^3$).

Theorem 21.

Let f be holomorphic in an open set Ω of the complex plane and let $z_0 \in \Omega$. Then there exists a neighborhood of z_0 on which f is univalent if and only if $f'(z_0) \neq 0$.

Proof:

Let us first assume that f is univalent on the open neighborhood Ω of z_0. Then $g = f^{-1}$ is holomorphic on $f(\Omega)$ by Theorem 20, so we may differentiate the identity $g(f(z)) = z$. We find via the chain rule that $g'(f(z_0))f'(z_0) = 1$, so $f'(z_0) \neq 0$.

Let us conversely assume that $f'(z_0) \neq 0$. Then the Jacobian of the mapping

$$f = u + iv \; : \; \mathbf{R}^2 \to \mathbf{R}^2$$

is $\neq 0$ (use the Cauchy-Riemann equations), so by the inverse function theorem f is injective on a neighborhood of z_0. $\qquad\square$

Remark: Theorem 21 extends to the case of holomorphic functions of several complex variables. An induction proof on the number of variables can be found in the note [Ro].

Section 4 Hadamard's gap theorem

As application of Theorem 8 we make an unnecessary, but nice, digression back to power series.

Theorem 22 (Hadamard's gap theorem).

Let $R > 0$ be the radius of convergence of the power series

$$f(z) = \sum_{k=0}^{\infty} a_k z^{n_k}$$

where $\{n_k \mid k = 1, 2, \cdots\}$ is a sequence of natural numbers satisfying $n_{k+1} \geq (1 + \theta)n_k$ for $k = 1, 2, \cdots$ for some $\theta > 0$.

Then $|z| = R$ is a natural boundary for f in the following sense: If there is a holomorphic extension of f to a connected open set $\Omega \supseteq B(0, R)$, then $\Omega = B(0, R)$.

Proof:

We assume for convenience that $R = 1$, and denote the extension of f by F. It suffices to derive a contradiction from the assumption $\Omega \neq B(0, 1)$.

Under that assumption $\Omega \cap S^1 \neq \emptyset$, say $1 \in \Omega$. Choose an integer $p \geq 1/\theta$ and consider for $w \in \mathbf{C}$ the convex combination

$$z = \frac{w^p + w^{p+1}}{2}$$

Section 4. Hadamard's gap theorem

We have

$$z \in B(0,1) \subseteq \Omega \text{ if } |w| \le 1 \text{ and } z \ne 1 \text{ , and}$$
$$z = 1 \text{ if } w = 1$$

so $z \in \Omega$ for all w in a neighborhood of $B[0,1]$.

The composite function

$$\phi(w) := F\left(\frac{w^p + w^{p+1}}{2}\right)$$

is therefore defined and holomorphic in that neighborhood of $B[0,1]$, and so the radius of convergence of ϕ's power series expansion around 0 is strictly bigger than 1. Let us find the power series expansion of ϕ: For $w \in B(0,1)$ we have

$$(*) \quad \sum_{k=0}^{\infty} a_k \left(\frac{w^k + w^{k+1}}{2}\right)^{n_k} = \sum_{k=0}^{\infty} \frac{a_k}{2^{n_k}} \left\{ w^{pn_k} + \binom{n}{1} w^{pn_k+1} + \cdots + w^{pn_k+n_k} \right\}$$

The exponent of the last term in the kth bracket, viz. $pn_k + n_k$, is strictly less than the exponent of the first term in the next bracket: Indeed, since $p \ge 1/\theta$, we see that

$$pn_{k+1} - (pn_k + n_k) \ge p(1 + \theta)n_k - (pn_k + n_k) \ge (p\theta - 1)n_k > 0$$

which means that no power of w occurs more than once.

We infer that the power series expansion of ϕ is gotten by simply omitting the brackets in (*): Indeed that power series converges, at least for $|w| < 1$, because

$$\sum_{k=0}^{\infty} \frac{|a_k|}{2^{n_k}} \left\{ |w|^{pn_k} + \binom{n}{1} |w|^{pn_k+1} + \cdots + |w|^{pn_k+n_k} \right\}$$

$$= \sum_{k=0}^{\infty} |a_k| \left(\frac{|w|^p + |w|^{p+1}}{2}\right)^{n_k} < \infty$$

since

$$\frac{|w|^p + |w|^{p+1}}{2} < 1$$

And by (*) the power series converges to ϕ.

As we mentioned above, the radius of convergence of ϕ is strictly bigger than 1. So its power series converges for some real number $r > 1$, implying that

$$\phi(r) = \sum_{k=0}^{\infty} a_k \left(\frac{r^k + r^{k+1}}{2}\right)^{n_k}$$

converges. But then the power series of f converges at

$$z = \frac{r^k + r^{k+1}}{2} > 1$$

and so the radius of convergence R of f is strictly bigger than 1.

That contradicts $R = 1$. $\qquad\square$

As an example we mention that the function

$$f(z) := \sum_{k=1}^{\infty} \frac{z^{n!}}{n!} \text{ for } z \in B(0,1)$$

extends to a continuous function on all of the complex plane, but that it cannot be prolonged to an analytic function on any domain bigger than $B(0,1)$.

Section 5 Exercises

1. In the proof of Weierstrass' theorem [Theorem 10] we used the fact that f analytic implies f holomorphic, i.e. Proposition II.4, so the remarks about power series following Weierstrass' theorem are really part of a circular argument. Repair the argument by showing directly from (**) in the proof of Weierstrass' theorem that f is holomorphic.

2. For which $n \in \mathbf{Z}$ will the function $z \to z^n$ possess a primitive 1) on all of $\mathbf{C}\backslash\{0\}$ 2) locally on $\mathbf{C}\backslash\{0\}$?

3. Let Ω be an open convex set, and let f be holomorphic on Ω with $\Re f' > 0$ on Ω. Show that f is univalent on Ω.

Hint: $f(b) - f(a) = (b - a)\int_0^1 f'(a + t(b - a))dt.$

4. Let $\gamma : [a, b] \to \mathbf{C}\backslash\{0\}$ be a path, and let f be continuous on γ^*. Prove the formula

$$\int_\gamma f(z)dz = -\int_{\gamma^{-1}} \frac{f(z^{-1})}{z^2}dz$$

in which the path $\gamma^{-1} : [a, b] \to \mathbf{C}$ is defined by $\gamma^{-1}(t) = 1/\gamma(t)$.

Show in particular that

$$\int_{|z|=r} f(z)dz = \int_{|z|=\frac{1}{r}} \frac{f(z^{-1})}{z^2}dz$$

5. Let $a \in \mathbf{C}$, $r > 0$ and $|a| \neq r$. Compute

$$\frac{1}{2\pi i}\int_{|z|=r} (z - a)^n dz \text{ for } n \in \mathbf{Z}$$

6. Let $\alpha \in \mathbf{C}$. Show that the principal branch of $(1 + z)^\alpha$, i.e. the branch with $(1 + 0)^\alpha = 1$ has the following power series expansion

$$(1 + z)^\alpha = \sum_{n=0}^{\infty} \binom{\alpha}{n} z^n \text{ for } |z| < 1$$

7. Let $\alpha \in C$. Is the following formula correct ?

$$\frac{d}{dz}(z^\alpha) = \alpha z^{\alpha-1}$$

8. Let f and g be continuous in the closed unit disc and analytic in the open disc. Write them as

$$f(z) = \sum_{n=0}^{\infty} a_n z^n \text{ and } g(z) = \sum_{n=0}^{\infty} b_n z^n \text{ for } |z| < 1$$

(α) Show for any $|z| < 1$ that

$$\sum_{n=0}^{\infty} a_n b_n z^{2n} = \frac{1}{2\pi i} \int_{|w|=1} f(zw)g\left(\frac{z}{w}\right)\frac{dw}{w}$$

(β) Show Parseval's identity

$$\sum_{n=0}^{\infty} a_n \overline{b_n} = \frac{1}{2\pi} \int_0^{2\pi} f\left(e^{i\theta}\right)\overline{g\left(e^{i\theta}\right)}d\theta$$

Hint : Show first for any $r \in]0,1[$ that

$$\sum_{n=0}^{\infty} |a_n|^2 r^{2n} = \frac{1}{2\pi} \int_0^{2\pi} \left|f\left(e^{i\theta}\right)\right|^2 d\theta$$

9. (L. Fejér and F. Riesz)

Let f be continuous on the closed unit disc and analytic in the interior. Show that

$$2\int_{-1}^{1} |f(x)|^2 dx \leq \int_0^{2\pi} \left|f\left(e^{i\theta}\right)\right|^2 d\theta$$

Hint : Apply the Cauchy integral theorem for the semi-circular disc to the function $z \to f(z)\overline{f(\overline{z})}$.

10. Let $t \in [-1,1]$

(α) Show that $1 - 2tz + z^2$ does not vanish in $|z| < 1$.

(β) Show that there exists a continuous square root of $z \to \left(1 - 2tz + z^2\right)^{-1}$ in $|z| < 1$ with the value 1 at $z = 0$. Call it $\left(1 - 2tz + z^2\right)^{-1/2}$.

(γ) Show that $z \to \left(1 - 2tz + z^2\right)^{-1/2}$ can be expanded in a power series

$$\left(1 - 2tz + z^2\right)^{-1/2} = \sum_{n=0}^{\infty} P_n(t)z^n \text{ converging for } |z| < 1$$

(δ) Show that the P_n are polynomials in t of degree n (They are the so-called *Legendre polynomials*).

(ϵ) Show that P_n is a solution of the equation

$$(1 - t^2)y'' - 2ty' + n(n+1)y = 0$$

Hint : Term by term integration yields the following explicit formula for P_n:

$$P_n(t) = \int\limits_{|z|=r} \left(1 - 2tz + z^2\right)^{-1/2} \frac{dz}{z^{n+1}}$$

11. (α) Show that $z \to \exp\left(-z^2 - 2zw\right)$ for any $w \in \mathbb{C}$ can be expanded in a power series

$$\exp\left(-z^2 - 2wz\right) = \sum_{n=0}^{\infty} H_n(w)\frac{z^n}{n!}$$

whose radius of convergence is ∞.

(β) Show that H_n is a polynomial in w of degree n (The so-called *Hermite polynomial of degree n*).

(γ) Show that the Hermite polynomials have the following properties :

$$H_n(w) = (-1)^n H_n(-w)$$
$$H'_{n+1} = -2(n+1)H_n$$
$$H''_n - 2wH'_n + 2nH_n = 0$$
$$H_n(w) = \exp\left(w^2\right)\left(\frac{d}{dw}\right)^n \left\{\exp\left(-w^2\right)\right\}$$

$$\int_{-\infty}^{\infty} H_n(x)H_m(x)\exp\left(-x^2\right)dx = 0 \text{ if } n \neq m$$

12. Consider the closed path γ composed of the four paths sketched below:

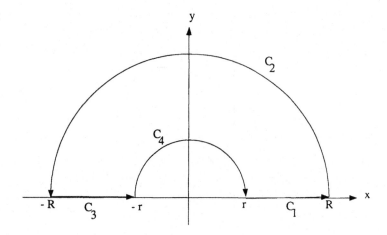

(α) Show that

$$\int_\gamma e^{iz}\frac{dz}{z} = 0$$

(β) Show that

$$\int_{C_2} e^{iz}\frac{dz}{z} \to 0 \ \text{ as } \ R \to \infty \text{ , and that}$$

$$\int_{C_4} e^{iz}\frac{dz}{z} \to -\pi i \ \text{ as } \ r \to 0$$

(γ) Show finally that

$$\int_0^R \frac{\sin x}{x}dx \to \frac{\pi}{2} \ \text{ as } \ R \to \infty$$

13. Let $0 < v < 1$. Let $A = 1 + e^{i(-\pi+v)}$ and $B = 1 + e^{i(\pi-v)}$. Let C and c denote the circle segments indicated on the figure.

63

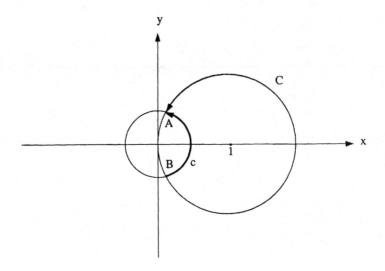

(α) Show that

$$\int_c \frac{Log\, z}{z-1} dz \to 0 \ \text{ as } \ v \to 0_+$$

(β) Show that

$$\int_{-\pi+v}^{\pi-v} Log\,|1+e^{it}|\, dt = \Im \left\{ \int_C \frac{Log\, z}{z-1} dz \right\} \to 0 \ \text{ as } \ v \to 0_+$$

(γ) Show that

$$\int_{\epsilon}^{\frac{\pi}{2}} Log(\sin\theta)\, d\theta \to -\frac{\pi}{2} Log\, 2 \ \text{ as } \ \epsilon \to 0_+$$

14. Does there exist a holomorphic function on $B(0,1)$ that in the points 1/2, 1/3, 1/4, 1/5, ... takes the following values :

(α) 1, 0, 1, 0, ...

(β) 1/2, 0, 1/4, 0, 1/6, 0, ...

(γ) 1/2, 1/4, 1/4, 1/6, 1/6, 1/8, 1/8, ...

(δ) 2/3, 3/4, 4/5, 5/6, ... ?

15. Let f be holomorphic and non-constant in an open connected subset Ω of the complex plane. Show that f is non-constant in any open subset of Ω.

16. We recall that

$$\sin z = \sum_{n=0}^{\infty} \frac{(-1)^n}{(2n+1)!} z^{2n+1} \quad \text{and} \quad \cos z = \sum_{n=0}^{\infty} \frac{(-1)^n}{(2n)!} z^{2n}$$

Show that $\sin^2 z + \cos^2 z = 1$ for all $z \in \mathbf{C}$.

17. Show that

$$\int_{-1}^{1} \frac{dt}{t-z} = Log\left\{ \frac{z-1}{z+1} \right\} \quad \text{for} \quad z \in \mathbf{C} \backslash [-1,1]$$

18. Let Ω_1 and Ω_2 be two open subsets of the complex plane such that $\Omega_1 \cap \Omega_2$ is connected. Assume that every holomorphic function on Ω_j has a primitive on Ω_j for $j = 1,2$.

Show that every function which is holomorphic on $\Omega_1 \cup \Omega_2$, has a primitive on $\Omega_1 \cup \Omega_2$.

19. Let f be holomorphic on an open subset Ω of the complex plane. Let g be a continuous k^{th} root of f on Ω.

Show that g, too, is holomorphic on Ω.

20. Let $g : B(a,r) \to B(g(a), R)$. Prove the following inequality which generalizes Schwarz' lemma:

$$|g(z+a) - g(a)| \le \frac{R}{r}|z| \quad \text{for all} \quad |z| < r$$

Derive Liouville's theorem from the inequality.

21. Show that the maximum of $|\sin z|$ on the square $[0, 2\pi] \times [0, 2\pi]$ equals $\cosh(2\pi)$ and is attained at $z = \frac{\pi}{2} + 2\pi i$.

22. Let f be an entire, non-constant function and consider the corresponding "analytic landscape"

$$\left\{ \left(x, y, |f(x+iy)|^2 \right) \in \mathbf{R}^3 \mid x, y \in \mathbf{R} \right\} \quad \text{in} \quad \mathbf{R}^3$$

Show that there are no highland lakes in such a landscape : When it rains the water collects in puddles around the zeros of f, not higher up, or runs off to infinity.

23. Prove the

Strong Maximum Principle:

Let Ω be a bounded, connected, open subset of the complex plane, and let f be a holomorphic, non-constant function on Ω. Let

$$M := \sup_{z \in \partial\Omega} \left\{ \limsup_{\zeta \to z, \zeta \in \Omega} |f(\zeta)| \right\}$$

Then $|f(z)| < M$ for all $z \in \Omega$.

24. Let Ω be a bounded open subset of the complex plane. Let $\{f_n\}$ be a sequence of functions which are continuous on $\overline{\Omega}$ and holomorphic on Ω. Assume that the sequence converges uniformly on the boundary of Ω.

Show that $\{f_n\}$ converges uniformly on $\overline{\Omega}$ towards a function which is continuous on $\overline{\Omega}$ and holomorphic on Ω.

25. Show that there exists a holomorphic function $A : B(0,1) \to \mathbf{C}$ with $A(0) = 0$ satisfying the identity $\sin(A(z)) = z$ for all $|z| < 1$.

Hint : Use Theorem 21 to get hold of A near 0. Differentiate the identity to get

$$A'(z) = \frac{1}{\sqrt{1 - z^2}}$$

26. Show that the Riemann ζ-function

$$\zeta(z) = \sum_{n=1}^{\infty} \frac{1}{n^z}$$

is holomorphic on the open half-plane $\{z \in \mathbf{C} \mid \Re z > 1\}$.

27. Show

Schwarz' Reflection Principle:

Let Ω be a connected open set which is symmetric with respect to the real axis. Put

$$\Omega_+ = \{z \in \Omega | \Im z > 0\}$$
$$\Omega_- = \{z \in \Omega | \Im z < 0\}$$
$$\Omega_0 = \{z \in \Omega | \Im z = 0\}$$

Let $f : \Omega_+ \cup \Omega_0 \to \mathbf{C}$ be continuous on $\Omega_+ \cup \Omega_0$, holomorphic on Ω_+ and such that $f(\Omega_0) \subseteq \mathbf{R}$.

Then there exists exactly one function $F : \Omega \to \mathbf{C}$ which is holomorphic on Ω and equal to f on Ω_+. It satisfies

$$F(z) = \overline{F(\overline{z})} \quad \text{for all } z \in \Omega$$

28. Assume that both $f \in C(\mathbf{R})$ and the function

$$x \to \int_{-\infty}^{\infty} e^{-ixt} f(t) dt \ , \ x \in \mathbf{R}$$

are compactly supported. Show that

$$\int\limits_{-\infty}^{\infty} e^{-ixt} f(t) dt = 0 \text{ for all } x \in \mathbf{R}$$

29. What is the maximum of the product of the four distances between a variable point in a square and the vertices of the square ?

30. Let $f : B(0,1) \to B(0,1)$ be an analytic map which is bijective and maps 0 into 0.

Show that f has the form $f(z) = cz$ for some constant $c \in S^1$.

31. Show that $|\sqrt{z}| = \sqrt{|z|}$. Why doesn't the function $z \to 1/\sqrt{z}$ constitute a counter example to the Removable Singularity Theorem ?

32. Let f be a non-constant entire function. Show that the range of f is dense in the complex plane. Hint: Assume not. Use Liouville.

33. We will say that an analytic function f, defined in $\Omega \subseteq \mathbf{C}$, is of *exponential type* if there is a $T \in \mathbf{R}$ and a constant $C > 0$ such that

$$|f(z)| \leq C e^{T|z|} \text{ for all } z \in \Omega$$

Prove

The Phragmén-Lindelöf Theorem:

If f is analytic and of exponential type in a sector Ω of angular opening strictly less than π, if f is also continuous in the closed sector $\overline{\Omega}$, and if $|f| \leq M$ for some constant M on the boundary of Ω, then $|f| \leq M$ throughout Ω.

34. Let f be holomorphic on an open subset Ω of the complex plane. Let $P, Q \in \Omega$ and let $\gamma_0, \gamma_1 : [a, b] \to \Omega$ be paths such that $\gamma_0(a) = \gamma_1(a) = P$ and $\gamma_0(b) = \gamma_1(b) = Q$.

Assume furthermore that γ_0 and γ_1 are fixed end point homotopic in Ω, i.e. there exists a continuous map $H : [a, b] \times [0, 1] \to \Omega$ such that

$$H(\cdot, 0) = \gamma_0 \;, \;\; H(\cdot, 1) = \gamma_1$$
$$H(a, s) = P \;, \;\; H(b, s) = Q \text{ for all } s \in [0, 1]$$

Show that

$$\int\limits_{\gamma_0} f(z) dz = \int\limits_{\gamma_1} f(z) dz$$

35. Let Ω be an open connected subset of **C**. Let $\{f_n\}$ be a sequence of continuous functions on Ω, converging locally uniformly to $f \in Hol(\Omega)$. Assume that none of the f_n vanishes at any point.

Show that f vanishes everywhere or nowhere.

36. Derive Brouwer's fixed point theorem (Theorem III.11) from Rouché's theorem.

37. In this exercise we will derive a formula which as special cases includes the *Poisson integral*

$$\int_0^\infty e^{-x^2} dx = \frac{\sqrt{\pi}}{2}$$

and the *Fresnel integrals*

$$\lim_{R \to \infty} \int_{-R}^R \cos\left(x^2\right) dx = \sqrt{\frac{\pi}{2}} \quad \text{and}$$

$$\lim_{R \to \infty} \int_{-R}^R \sin\left(x^2\right) dx = \sqrt{\frac{\pi}{2}}$$

To that purpose we consider the meromorphic function

$$f(z) := \frac{e^{iz^2}}{e^{-\sqrt{2\pi z}} - 1} \quad , \ z \in \mathbf{C}$$

(a) Show, to ease your computations below, that

$$f(z) + f(-z) = -e^{iz^2} \quad \text{and}$$
$$f\left(i\sqrt{\frac{\pi}{2}} + z\right) + f\left(i\sqrt{\frac{\pi}{2}} - z\right) = i e^{iz^2}$$

Consider for each $R > 0$ and fixed $\alpha \in [0, \frac{\pi}{2}[$ the following parallelogram A, B, C, D , in which

$$A = -B = -Re^{i\alpha} \ , \ C = B + i\sqrt{\frac{\pi}{2}} \ , \ D = A + i\sqrt{\frac{\pi}{2}}$$
$$E = i\sqrt{\frac{\pi}{2}} \ , \ F = -G = re^{i\alpha}$$

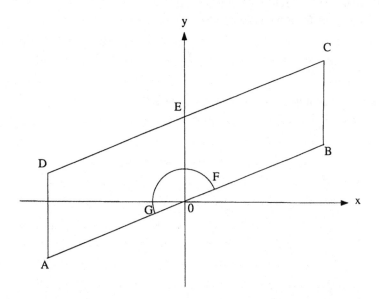

Note that f is holomorphic on the interior of the parallelogram and on its boundary only has a singularity at $z = 0$.

(b) Show that

$$\int_{[B,C]} f(z)dz \to 0 \quad \text{and} \quad \int_{[A,D]} f(z)dz \to 0 \quad \text{as} \ R \to \infty$$

(c) Show that

$$\int_{[C,D]} f(z)dz = -ie^{i\alpha} \int_0^R \exp\left\{it^2 e^{2i\alpha}\right\} dt$$

and that

$$\int_{[A,G]} f(z)dz + \int_{[F,B]} f(z)dz = -e^{i\alpha} \int_r^R \exp\left\{it^2 e^{2i\alpha}\right\} dt$$

Hint: Use (a).

(d) Show that the integral of f along the small half circle on the figure from G to F as $r \to 0_+$ converges to $i\sqrt{\frac{\pi}{2}}$.

(e) Show that

$$\lim_{R\to\infty} \int_0^R \exp\left\{it^2 e^{2i\alpha}\right\} dt = \frac{1+i}{2}\sqrt{\frac{\pi}{2}}e^{-i\alpha}$$

(f) Derive the Poisson and Fresnel integrals.

(g) Show the following more compact form of the result of (e):

$$\lim_{R \to \infty} \int_0^R e^{i\pi a x^2/2} dx = \frac{1+i}{2} \frac{1}{\sqrt{a}} \quad \text{for } a \neq 0 \text{ , } \Im a \geq 0$$

where the principal value of $\sqrt{a}$ is taken on the right.

The exercise is adapted from the paper [Ra].

Chapter 5 Global Theory

Section 1 The global Cauchy integral theorem

In this section we will state and prove a general version of the Cauchy integral formula and the Cauchy integral theorem.

Theorem IV.6 states the Cauchy integral formula for a closed path in a starshaped domain. We would of course like to remove the condition of starshapedness, but we must be careful, for if we do so then some kind of restriction on the paths in question is necessary:

Take as domain the annulus $A := \left\{ z \in \mathbf{C} \mid \frac{1}{2} < |z| < 1 \right\}$ and consider functions which are holomorphic in an open set containing the closure of A. It is simply not true that

$$f(z) = \frac{1}{2\pi i} \int\limits_{|\zeta|=1} \frac{f(\zeta)}{\zeta - z} d\zeta \ \text{ for } \ z \in A$$

as the function $f(z) = z^{-1}$ demonstrates.

On the positive side there exists a version of the Cauchy integral formula in A, viz.

$$f(z) = \frac{1}{2\pi i} \int\limits_{|\zeta|=1} \frac{f(\zeta)}{\zeta - z} d\zeta - \frac{1}{2\pi i} \int\limits_{|\zeta|=\frac{1}{2}} \frac{f(\zeta)}{\zeta - z} d\zeta \ \text{ for } \ z \in A$$

(Exercise: Derive it from the Cauchy integral formula for starshaped domains by inserting suitable auxiliary segments).

In the case of a starshaped domain Ω the Cauchy theorem states that

$$(*) \quad \int\limits_{\gamma} f(z) dz = 0$$

for any closed path γ in Ω and any function f which is holomorphic in Ω. Let now Ω be any open subset of the complex plane, and let γ be a closed path in Ω. If $(*)$ holds for all $f \in Hol(\Omega)$, then in particular

$$Ind_\gamma(z) = \frac{1}{2\pi i} \int\limits_{\gamma} \frac{d\zeta}{\zeta - z} = 0 \ \text{ for all } \ z \in \mathbf{C} \backslash \Omega$$

Our general version of the Cauchy integral theorem is that the converse holds. The crucial necessary condition on the path γ, i.e. that $Ind_\gamma(z) = 0$ for all $z \in \mathbf{C} \backslash \Omega$, means intuitively that γ does not circle any "hole" in Ω.

71

Theorem 1 (The global Cauchy theorem and integral formula).

Let γ be a closed path in an open subset Ω of the complex plane such that $Ind_\gamma(z) = 0$ for all $z \in \mathbf{C}\backslash\Omega$.

Then we have for any $f \in Hol(\Omega)$ that

$$(\alpha) \quad \int_\gamma f(z)dz = 0 \ , \ and$$

$$(\beta) \quad f(z)\,Ind_\gamma(z) = \frac{1}{2\pi i}\int_\gamma \frac{f(\zeta)}{\zeta - z}d\zeta \ \ for \ all \ z \in \Omega\backslash\gamma^*$$

Example 2.

Here is a drawing of a path that satisfies the condition of Theorem 1, but which nevertheless is not null-homotopic $[\Omega = C\backslash\{0,1\}]$:

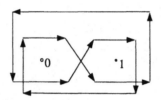

Proof of Theorem 1: (due to [Di]). (α) follows from (β) when we replace f by $\zeta \to (\zeta - z)f(\zeta)$, so from now on we may concentrate on (β).

By the formula for the index (Theorem III.17(d)) (β) is equivalent to

$$\frac{1}{2\pi i}\int_\gamma \frac{f(z) - f(w)}{z - w}dw = 0 \ \ \text{for all} \ z \in \Omega\backslash\gamma^*$$

Introducing the function $h : \Omega \times \Omega \to \mathbf{C}$ given by

$$h(z,w) := \begin{cases} \frac{f(z)-f(w)}{z-w} & \text{for } z \neq w \\ f'(w) & \text{for } z = w \end{cases}$$

it thus suffices to prove that

$$g(z) := \frac{1}{2\pi i}\int_\gamma h(z,w)dw = 0 \ \ \text{for all} \ z \in \Omega$$

We start by noting some of the properties of h and g. h is continuous in all of $\Omega \times \Omega$: This is obvious off the diagonal in $\Omega \times \Omega$. To prove that h is continuous on the diagonal as well we compute

$$f(z) - f(w) = \int\limits_0^1 \frac{d}{dt}\{f(w + t(z - w))\}dt = (z - w) \int\limits_0^1 f'(w + t(z - w))dt$$

which implies that h near the diagonal can be expressed as

$$h(z, w) = \int\limits_0^1 f'(w + t(z - w))dt$$

a formula that also holds on the diagonal. This formula proves the continuity of h on the diagonal.

So g makes sense and is continuous in Ω. By help of Morera's theorem (Theorem IV.9) we will prove that g is even holomorphic in Ω :

If Δ is a triangle in Ω then

$$\int\limits_{\partial\Delta} g = \frac{1}{2\pi i} \int\limits_{\partial\Delta} \int\limits_\gamma h(z, w)dwdz = \frac{1}{2\pi i} \int\limits_\gamma \int\limits_{\partial\Delta} h(z, w)dzdw$$

Now, the function $z \to h(z, w)$ is holomorphic in Ω for fixed w [Example IV.4], so according to the Cauchy-Goursat theorem

$$\int\limits_{\partial\Delta} h(z, w)dz = 0 \text{ , and hence } \int\limits_{\partial\Delta} g = 0$$

so according to Morera g is holomorphic in Ω.

By our assumption on the path γ the set

$$\Omega_0 := \{z \in \mathbf{C}\backslash\gamma^* \mid Ind_\gamma(z) = 0\}$$

is an open subset of $\mathbf{C}$ satisfying $\Omega \cup \Omega_0 = \mathbf{C}$.

For $z \in \Omega \cap \Omega_0$ we get:

$$g(z) = \frac{1}{2\pi i} \int\limits_\gamma h(z, w)dw = \frac{1}{2\pi i} \int\limits_\gamma \frac{f(z)}{z - w}dw - \frac{1}{2\pi i} \int\limits_\gamma \frac{f(w)}{z - w}dw$$

$$= -f(z)\, Ind_\gamma(z) + \frac{1}{2\pi i} \int\limits_\gamma \frac{f(w)}{w - z}dw = \frac{1}{2\pi i} \int\limits_\gamma \frac{f(w)}{w - z}dw$$

The function

$$g_0(z) := \frac{1}{2\pi i} \int\limits_\gamma \frac{f(w)}{w - z}dw \text{ for } z \in \Omega_0$$

is holomorphic on Ω_0 (Lemma II.13); the computation just made shows that g and g_0 patch together over $\Omega \cap \Omega_0$ to an unambiguously defined function G on $\Omega \cup \Omega_0 = \mathbf{C}$, i.e. G given by

$$G(z) := \begin{cases} g(z) & \text{for } z \in \Omega \\ g_0(z) & \text{for } z \in \Omega_0 \end{cases}$$

is entire.

The unbounded component of γ^* is by Theorem III.17(c) contained in Ω_0, so for z out there we can estimate G as follows :

$$|G(z)| = |g_0(z)| = \frac{1}{2\pi} \left| \int_\gamma \frac{f(w)}{w-z} dw \right| \leq \frac{1}{2\pi} \frac{\sup\{|f(w)| \mid \omega \in \gamma^*\}}{dist(z, \gamma^*)} L(\gamma)$$

which shows that $G(z) \to 0$ as $z \to \infty$. But then G vanishes identically by Liouville's theorem (Theorem I.11). In particular g is identically 0 on Ω, and that fact is exactly the contents of (β). $\qquad\square$

An inspection of the proof reveals that it applies to the case of several paths, not just one. This comes in handy for, say, an annulus as mentioned in the introduction of this section. The precise statement is :

Theorem 3 (The global Cauchy theorem and integral formula).

Let $\gamma_1, \gamma_2, \cdots, \gamma_n$ be closed paths in an open subset Ω of the complex plane such that

$$\sum_{j=1}^n Ind_{\gamma_j}(z) = 0 \ \text{for all } z \in \mathbf{C} \backslash \Omega$$

Then we have for any $f \in Hol(\Omega)$ that

$$(\alpha) \ \sum_{j=1}^n \int_{\gamma_j} f(z)dz \ = 0 \ , \ and$$

$$(\beta) \ f(z) \sum_{j=1}^n Ind_{\gamma_j}(z) = \sum_{j=1}^n \frac{1}{2\pi i} \int_{\gamma_j} \frac{f(w)}{w-z} dw$$

$$\text{for all } z \in \Omega \backslash \{\gamma_1^* \cup \cdots \cup \gamma_n^*\}$$

Proof: Replace in the proof of Theorem 1 above

$$g \quad \text{by} \quad \sum_{j=1}^n \frac{1}{2\pi i} \int_{\gamma_j} h(z, w)dw$$

$$\Omega_0 \quad \text{by} \quad \left\{ z \in \mathbf{C} \backslash \{\gamma_1^* \cup \cdots \cup \gamma_n^*\} \mid \sum_{j=1}^n Ind_{\gamma_j}(z) = 0 \right\}$$

$$g_0 \quad \text{by} \quad \frac{1}{2\pi i} \sum_{j=1}^n \int_{\gamma_j} \frac{f(w)}{w-z} dw$$

The details of the easy modification of the proof of Theorem 1 are left to the reader. □

Corollary 4.

Let f be holomorphic in an open subset Ω of the complex plane. Let $\gamma_1, \gamma_2, \cdots, \gamma_N$ and $\sigma_1, \sigma_2, \cdots, \sigma_M$ be closed paths in Ω and let $n_1, n_2, \cdots, n_N$ and $m_1, m_2, \cdots, m_M$ be integers such that

$$\sum_{j=1}^{N} n_j \, Ind_{\gamma_j}(z) = \sum_{l=1}^{M} m_l \, Ind_{\sigma_l}(z) \ \text{ for all } \ z \in \mathbf{C}\backslash\Omega$$

Then

$$\sum_{j=1}^{N} n_j \int_{\gamma_j} f = \sum_{l=1}^{M} m_l \int_{\sigma_l} f$$

Corollary 5.

Let f be holomorphic in an open subset Ω of the complex plane. Let γ and ρ be two closed paths in Ω, which are homotopic in Ω. Then

$$\int_{\gamma} f = \int_{\rho} f$$

In particular, if γ is null-homotopic in Ω, then $\int_{\gamma} f = 0$.

Section 2 Simply connected sets

The topological concept of simple connectivity is intimately related to homotopy (Definition III.16).

Definition 6.

A topological space is said to be simply connected, *if each closed curve in it is null-homotopic.*

It is obvious that any starshaped space is simply connected. $\mathbf{R}^3\backslash\{0\}$ is an example of a topological space which is simply connected, but not starshaped. It is intuitively obvious that $\mathbf{R}^2\backslash\{0\}$ is connected, but not simply connected (For a rigorous proof see the remark following Theorem 9). $B(0,1) \cup B(3,1)$ is simply connected, but not connected.

Theorem 7.

Let Ω be a simply connected open subset of **C**, *let γ be a closed path in Ω and let $f \in Hol(\Omega)$. Then $\int_\gamma f = 0$ and*

$$f(z) \, Ind_\gamma(z) = \frac{1}{2\pi i} \int_\gamma \frac{f(w)}{w-z} dw \ \text{ for all } \ z \in \Omega \backslash \gamma^*$$

Proof: Obvious from Corollary 5. $\qquad\qquad\qquad\qquad\qquad\qquad\qquad\square$

Theorem 8.

Let Ω be a simply connected open subset of **C**. *Then any $f \in Hol(\Omega)$ has a (holomorphic) primitive on Ω.*

Proof: It suffices to construct a primitive on each of the connected components of Ω, so we may as well from the start assume that Ω is connected. Choose $z_0 \in \Omega$ and let γ_z be a path in Ω from z_0 to $z \in \Omega$. Then

$$F(z) \ := \ \int_{\gamma_z} f(w) dw$$

is according to Theorem 7 independent of the choice of path from z_0 to z, so F is well defined on Ω. Proceeding as in the proof of Lemma II.12 we see that $F' = f$. $\qquad\square$

We can now as promised extend Theorem III.21 and III.10 to simply connected sets.

Theorem 9.

Let Ω be a simply connected open subset of **C** *and let f be a never vanishing holomorphic function on Ω. Then f has a holomorphic logarithm and a holomorphic square root on Ω.*

Proof: It suffices to construct the logarithm and the square root in each of the connected components of Ω, so we may as well assume that Ω is connected. Let F be a primitive of f'/f in Ω; such one exists by Theorem 8. The function fe^{-F} is constant because

$$\left(fe^{-F}\right)' = f'e^{-F} - fF'e^{-F} = \left(f' - f\frac{f'}{f}\right)e^{-F} = 0$$

so $f = ce^F$ for some constant c. Writing $c = e^w$ we see that $F - w$ is a holomorphic logarithm of f.

A holomorphic square root is $\exp\left[(F - w/2)\right]$. $\qquad\qquad\qquad\qquad\qquad\square$

In passing we note that this gives us a rigorous proof that $\mathbf{R}^2 \backslash \{0\}$ is not simply connected : Combine Example III.22(b) and Theorem 9.

For other results on simply connected sets see the discussion around the Riemann mapping theorem (Theorem VIII.20), Chapter VIII.§4, Remark IX.11 and Exercise IX.20.

Section 3 Exercises

1. Consider the path γ in Example 2. What is

$$\int_\gamma \frac{dz}{z} \ ?$$

$$\int_\gamma \frac{dz}{z - \epsilon} \quad \text{when } \epsilon > 0 \text{ is small ?}$$

$$\int_\gamma \frac{e^z - e^{-z}}{z^6} dz \ ?$$

2. Let f be holomorphic in $\mathbf{C} \backslash \{0\}$. Show that

$$\int_{|z|=R} f = \int_{|z|=1} f \quad \text{for all } R > 0$$

3. Show that the function $1 - z^2$ has a holomorphic square root in any simply connected open subset of the complex plane which does not contain $\{-1, 1\}$. Find the possible values of

$$\int_\gamma \frac{dz}{\sqrt{1 - z^2}}$$

when γ is a closed path in such a domain.

4. Let Ω be an open, connected and simply connected subset of the complex plane, and let $z_0, a_1, a_2, \cdots, a_N \in \Omega$ be distinct points. Choose $\epsilon > 0$ so small that the closed balls $B[a_1, \epsilon], \cdots, B[a_N, \epsilon]$ are contained in Ω and so that $a_j \notin B[a_k, \epsilon]$ when $j \neq k$. For $h \in Hol(\Omega \backslash \{a_1, a_2, \cdots, a_N\})$ we introduce the constants

$$c_j = \frac{1}{2\pi i} \int_{|w - a_j| = \epsilon} h(w) dw \in \mathbf{C}$$

Let $\gamma = \gamma_z$ be any path in $\Omega \backslash \{a_1, a_2, \cdots, a_N\}$ from z_0 to z. Show that the following function of z :

$$\int_\gamma \left\{ h(w) - \frac{c_1}{w - a_1} - \cdots - \frac{c_N}{w - a_N} \right\} dw$$

is independent of the choice of path from z_0 to z.

5. Theorem 9 deals with holomorphic functions, but the result is really a topological one. The purpose of this exercise is to establish the following topological version of Theorem 9:

Theorem 10.

Let Ω be a simply connected open subset of the complex plane and let $f : \Omega \to \mathbb{C}\backslash\{0\}$ be continuous. Then f has a continuous logarithm on Ω.

It suffices to prove the theorem for each of the connected components of Ω, so we may as well assume that Ω is connected.

(α) Prove that $Ind_{f\circ\gamma}(0) = 0$ for all closed curves γ in Ω.

(β) Let $\gamma_1, \gamma_2 : [0,1] \to \Omega$ be continuous curves such that $\gamma_1(0) = \gamma_2(0)$ and $\gamma_1(1) = \gamma_2(1)$. Let $\phi_1, \phi_2 : [0,1] \to \mathbb{C}$ be continuous logarithms of $f \circ \gamma_1$ and $f \circ \gamma_2$ respectively, starting at the same point $\phi_1(0) = \phi_2(0)$.

Show that $\phi_1(1) = \phi_2(1)$.

(γ) Fix $z_0 \in \Omega$ and $w_0 \in \mathbb{C}$ such that $\exp w_0 = f(z_0)$. Find for any $z \in \Omega$ a curve γ such that $\gamma(0) = z_0$ and $\gamma(1) = z$ and take a continuous logarithm ϕ of $f \circ \gamma$ such that $\phi(0) = w_0$.

Show that $\phi(1)$ does not depend on the choice of the curve γ, so that we unambiguously may define $F(z) := \gamma(1)$.

(δ) Show that F is a logarithm of f.

(ϵ) Show that $F \circ \gamma$ is continuous for any curve γ from (γ).

(ζ) Show that F is a continuous logarithm of f.   □

Chapter 6 Isolated Singularities

We begin the present chapter with a study of functions which are holomorphic in an annulus; we show that they can be expanded in Laurent series in much the same spirit as functions which are holomorphic in a disc can be expanded in power series. The special case of a holomorphic function, defined in a punctured disc is the topic of the remainder of the chapter. The center of the disc is in that case said to be an isolated singularity of the function. We classify isolated singularities into removable singularities, poles and essential singularities. We finally prove the Residue Theorem and use it to evaluate definite integrals of various types; this is certainly one of the high points of any introductory course on complex analysis.

A deeper study of essential singularities can be found in the next chapter in which the two Picard theorems are derived.

Section 1 Laurent series

Definition 1.

A Laurent series *is a series of the form*

$$(1) \quad \sum_{n=-\infty}^{\infty} a_n z^n$$

where the coefficients a_n for $n \in \mathbf{Z}$ are complex numbers, and where z is a complex number different from 0.

If the two series

$$(2) \quad \sum_{n=-\infty}^{-1} a_n z^n = \sum_{m=1}^{\infty} a_{-m} \left(\frac{1}{z}\right)^m \quad \text{and}$$

$$(3) \quad \sum_{n=0}^{\infty} a_n z^n$$

both are convergent with sums S_- and S_+ respectively, then we will say that the Laurent series (1) converges and that its sum is $S_- + S_+$. If at least one of the series (2) and (3) diverges, then we will say that the Laurent series (1) diverges.

The concepts of absolute convergence and uniform convergence are defined similarly.

By definition the convergence of the Laurent series (1) means convergence of the two series (2) and (3) which are power series in $1/z$ and z respectively. So it is to be expected that results, analogous to those for power series, hold for Laurent series.

Theorem 2.

Let
$$r := \limsup_{m \to \infty} |a_{-m}|^{\frac{1}{m}} \ \text{ and } \ R := \frac{1}{\limsup_{n \to \infty} |a_n|^{\frac{1}{n}}}$$

Assume that $0 \leq r < R \leq \infty$, and let A denote the annulus

$$A := \{z \in \mathbf{C} \mid r < |z| < R\}$$

The Laurent series (1) converges absolutely at each point of A, and uniformly on any compact subset of A. When $|z| < r$ or $|z| > R$ the Laurent series (1) diverges.

The function f, given by

$$f(z) := \sum_{n=-\infty}^{\infty} a_n z^n \ \text{ for } \ z \in A$$

is holomorphic in A, and its derivative can be found by term by term differentiation, i.e.

$$f'(z) = \sum_{n=-\infty}^{\infty} n a_n z^{n-1} \ \text{ for } \ z \in A$$

We know that a function which is holomorphic in an open disc may be expanded in a power series there. We will now show the corresponding result for a function which is holomorphic in an annulus.

Theorem 3.

Let f be holomorphic in an annulus $A := \{z \in \mathbf{C} \mid r < |z| < R\}$ where $0 \leq r < R \leq \infty$.

Then there exists exactly one Laurent series (1) such that it converges at each point of A and such that

$$f(z) = \sum_{n=-\infty}^{\infty} a_n z^n \ \text{ for } \ z \in A$$

The coefficients of this Laurent series can be found from f as follows

$$a_n = \frac{1}{2\pi i} \int_{|w|=\rho} \frac{f(w)}{w^{n+1}} dw \ \text{ for } \ n \in \mathbf{Z}$$

where ρ is arbitrary in the open interval $]r, R[$.

Proof: Assume that f is the sum of a Laurent series (1) in A. By the previous theorem the series converges uniformly on the circle $|w| = \rho$, so

$$\frac{1}{2\pi i} \int_{|w|=\rho} \frac{f(w)}{w^{n+1}} dw = \frac{1}{2\pi i} \sum_{m=-\infty}^{\infty} a_m \int_{|w|=\rho} \frac{dw}{w^{n-m+1}} = \sum_{m=-\infty}^{\infty} a_m \delta_{mn} = a_n$$

This proves the uniqueness of a possible Laurent expansion and that the coefficients will be the ones indicated in the statement of the theorem. It is left to show that f is the sum of a Laurent series in A.

Choose r' and R' such that $r < r' < R' < R$. If $r' < |z| < R'$ we have from the global Cauchy formula (Theorem V.3) that

$$f(z) = \frac{1}{2\pi i} \int\limits_{|w|=R'} \frac{f(w)}{w-z} dw - \frac{1}{2\pi i} \int\limits_{|w|=r'} \frac{f(w)}{w-z} dw$$

The first integral, i.e. the function

$$G(z) = \frac{1}{2\pi i} \int\limits_{|w|=R'} \frac{f(w)}{w-z} dw$$

is holomorphic in $\{z \in \mathbf{C} \mid |z| \neq R'\}$, hence in particular in the disc $B(0, R')$, and so it has a power series expansion

$$G(z) = \sum_{m=0}^{\infty} b_m z^m \text{ for } |z| < R'$$

The second integral, i.e. the function

$$g(z) = -\frac{1}{2\pi i} \int\limits_{|w|=r'} \frac{f(w)}{w-z} dw$$

may for $|z| > r'$ be treated as follows:

$$g(z) = -\frac{1}{2\pi i} \int\limits_{|w|=r'} \frac{f(w)}{w-z} dw = \frac{1}{2\pi i} \frac{1}{z} \int\limits_{|w|=r'} \frac{f(w)}{1-\frac{w}{z}} dw$$

$$= \frac{1}{2\pi i} \frac{1}{z} \int\limits_{|w|=r'} f(w) \sum_{n=0}^{\infty} \frac{w^n}{z^n} dw$$

$$= \frac{1}{2\pi i} \frac{1}{z} \sum_{n=0}^{\infty} \frac{1}{z^n} \int\limits_{|w|=r'} f(w) w^n dw$$

which exhibits a Laurent expansion of g. We conclude that $f = G + g$ has a Laurent expansion in the annulus $\{z \in \mathbf{C} \mid r' < |z| < R'\}$. But by uniqueness the Laurent series will be the same irregardless of the values of r' and R', so it converges in all points of A and has the sum f there. $\qquad\qquad \square$

Laurent series constitute an important tool in digital signal processing in form of the Z-transform: A signal is a two-sided sequence $\{a_n\}_{n \in \mathbf{Z}}$ of complex numbers; its

Z-transform is by definition the sum of the Laurent series (1). This is a generalization of the corresponding Fourier series

$$\sum_{n=-\infty}^{\infty} a_n e^{-in\theta}$$

but the Laurent series may converge even if the Fourier series does not. The Z-transform has many pleasant properties: E.g. the Z-transform of a convolution is the product of the Z-transforms. For more information on the Z-transform and its applications the reader may look in any book on digital signal processing, say the classic monograph [OS].

Section 2 The classification of isolated singularities

Definition 4.

If a function f is holomorphic on an open set of the form $B(a,r)\backslash\{0\}$, we say that a is an isolated singularity *of f.*

As examples of isolated singularities the reader can have the following three functions in mind :

$$\frac{\sin z}{z} \; , \; \; \frac{\cos z}{z} \; \; \text{and} \; \; \exp\left(\frac{1}{z}\right)$$

Each of them has an isolated singularity at the point $z = 0$.

The classification of an isolated singularity of f is based on the behavior of f at the singularity a. There are three mutually excluding cases:

In the first case f remains bounded near a. This case is taken care of by the Removable Singularity Theorem (Theorem IV.3) : f extends across a to an analytic function on all of $B(a,r)$. We say that a is a *removable singularity* of f. An example: $z = 0$ is a removable singularity of the function $\frac{\sin z}{z}$. We will always let f denote also the extended function, even if we don't say so explicitly. So for example we give the function $\frac{\sin z}{z}$ the value 1 at $z = 0$.

In the second case f has a *pole* at $z = a$ meaning by definition that $|f(z)| \to \infty$ as $z \to a$. Here the function $g(z) := 1/f(z)$ has a removable singularity at $z = a$ where it takes the value 0, so we can write it in the form

$$g(z) = (z - a)^n h(z)$$

where $n \in \{1, 2, \cdots\}$, and where h is analytic near a and $h(a) \neq 0$. Thus $(z - a)^n f(z)$ remains bounded near $z = a$. The smallest integer $n > 0$ such that $(z - a)^n f(z)$ remains bounded near a is called the *order of the pole*. It equals the order of the zero of $1/f$. If a pole has order 1 we call it a *simple pole*. An example of a simple pole is $\frac{\cos z}{z}$. A more general example of a simple pole is a quotient of the form $f(z) = g(z)/h(z)$ where g and h are holomorphic near $z = a$, $h(a) = 0$, $h'(a) \neq 0$ and $g(a) \neq 0$.

The third and final case is the one in which $\lim |f(z)|$ as $z \to a$ does not exist in $\mathbf{R} \cup \{\infty\}$. We say that f has an *essential singularity* at a. An example is $f(z) = \exp\left(z^{-1}\right)$.

The beautiful relation between the coefficients of the Laurent expansion of f and the behavior of f at the isolated singularity is taken up in Exercise 1 below.

Definition 5.

Let a be an isolated singularity of the holomorphic function $f : B(a,r)\backslash\{a\} \to \mathbf{C}$. By the residue of f at a we mean the complex number

$$Res(f;a) := \frac{1}{2\pi i} \int\limits_{|z-a|=\epsilon} f(z)dz$$

where ϵ is any number in the interval $]0, r[$. By Corollary V.5 the residue of f at a is independent of the choice of ϵ.

The next sections will show that residues are important for the evaluation of definite integrals. So we must be able to compute residues. The simplest instance occurs when f has a removable singularity at a. Here $Res(f;a) = 0$ by the Cauchy integral theorem (Theorem IV.5). Another particularly simple case - which even occurs quite often - is the case of a simple pole. This is singled out as a special instance in the following recipe for computing the residue at a pole.

Proposition 6.

(a) If f has a pole of order $n < \infty$ at $z = a$ then the function $(z - a)^n f(z)$ has a removable singularity at $z = a$, and

$$Res(f;a) = \frac{1}{(n-1)!}\left(\frac{d}{dz}\right)^{n-1}\left\{(z-a)^n f(z)\right\}|_{z=a}$$

(b) In particular, if f has a simple pole at $z = a$ then

$$Res(f;a) = \left\{(z-a)f(z)\right\}_{z=a}$$

(c) If f and g are holomorphic near $z = a$, $g(a) = 0$ and $g'(a) \neq 0$, then

$$Res\left(\frac{f}{g};a\right) = \frac{f(a)}{g'(a)}$$

Proof:

We prove only (b) and (c) and leave the easy generalization (a) to the reader.

(b) The definition of a simple pole implies that $g(z) := (z - a)f(z)$ remains bounded near $z = a$, so that g has a removable singularity at $z = a$. Now, let's write

$$f(z) = \frac{g(z)}{z-a} = \frac{g(z) - g(a)}{z-a} + \frac{g(a)}{z-a}$$

and observe that the first term on the right hand side is holomorphic across the singularity $z = a$ by the Removable Singularity Theorem. So

$$\int\limits_{|z-a|=\epsilon} f(z)dz = \int\limits_{|z-a|=\epsilon} \frac{g(z) - g(a)}{z - a}dz + g(a) \int\limits_{|z-a|=\epsilon} \frac{dz}{z - a} = 0 + g(a)2\pi i$$

proving (b).

(c) Since f/g has a simple pole (if $f(a) \neq 0$) or a removable singularity (if $f(a) = 0$) at $z = a$ we get from (b) that

$$Res\left(\frac{f}{g}; a\right) = \lim_{z \to a} \left\{ (z - a)\frac{f(z)}{g(z)} \right\}$$

$$= \lim_{z \to a} \frac{f(z)}{(g(z) - g(a))/(z - a)} = \frac{f(a)}{g'(a)}$$

$\square$

As a final example of how to compute residues we consider the function $\exp\left(z^{-1}\right)$ which has an essential singularity at $z = 0$. Since the series

$$\exp\left(\frac{1}{z}\right) = \sum_{n=0}^{\infty} \frac{1}{n!} z^{-n}$$

converges uniformly on compact subsets of $\mathbb{C}\backslash\{0\}$ we can integrate term by term and find

$$Res\left(\exp\left(\frac{1}{z}\right); 0\right) = \frac{1}{2\pi i} \sum_{n=0}^{\infty} \frac{1}{n!} \int\limits_{|z|=\epsilon} z^{-n}dz = \sum_{n=0}^{\infty} \frac{1}{n!}\delta_{1n} = 1$$

Of course, by the same method we see more generally that any function f which around $z = a$ has a Laurent expansion

$$f(z) = \sum_{n=-\infty}^{\infty} a_n(z - a)^n$$

has $Res(f; a) = a_{-1}$.

Section 3 The residue theorem

The statement.

The residue theorem generalizes both the Cauchy integral theorem and the Cauchy integral formula. Our version of it runs as follows:

84

Theorem 7 (The Cauchy residue theorem).

Let Ω be an open subset of the complex plane and let γ be a closed path in Ω such that $Ind_\gamma(z) = 0$ for all $z \in \mathbb{C} \backslash \Omega$. Let f be holomorphic in Ω except for finitely many isolated singularities $a_1, a_2, \cdots, a_N \in \Omega \backslash \gamma^*$. Then

$$\frac{1}{2\pi i} \int_\gamma f(z)dz = \sum_{k=1}^{N} Ind_\gamma(a_k)\, Res(f; a_k)$$

The hypothesis on γ is clearly satisfied if γ is homotopic in Ω to a constant curve, say if Ω is simply connected. In most examples γ will be a simple closed path traversed counterclockwise, and $Ind_\gamma(z)$ will be 1 or 0 according to whether z is inside or outside γ.

Proof: Choose closed paths $\gamma_1, \gamma_2, \cdots, \gamma_N$ in Ω such that each γ_j circles around the singularity a_j once and does not contain any of the other singularities of f. More precisely choose them such that

$$Ind_{\gamma_j}(z) \ = 0 \ \text{for all} \ z \in \mathbb{C} \backslash \Omega \ \text{and} \ j = 1, 2, \cdots, N$$
and
$$Ind_{\gamma_j}(a_k) = \delta_{jk} \ \text{for} \ j, k = 1, 2, \cdots, N$$

Then

$$Ind_\gamma(z) = \sum_{j=1}^{N} Ind_\gamma(a_j)\, Ind_{\gamma_j}(z) \ \text{for all} \ z \in \mathbb{C} \backslash (\Omega \backslash \{a_1, a_2, \cdots, a_N\})$$

so by the Global Cauchy Theorem V.3 (or better its Corollary V.4)

$$\int_\gamma f = \sum_{j=1}^{N} Ind_\gamma(a_j) \int_{\gamma_j} f$$

Dividing through by $2\pi i$ we get the Residue Theorem. $\qquad\qquad \square$

The Cauchy Residue theorem can be used to evaluate definite integrals. How successful one is depends largely on correct choice of contour and representative function. The techniques will be illustrated below by various examples that will provide us with recipes for evaluation of definite integrals of different types.

Example A.

An integral of the form

$$\int_0^{2\pi} F(\cos\theta, \sin\theta)d\theta$$

is by the substitution $z = \exp{(i\theta)}$ converted into a complex line integral as follows:

$$\int\limits_{0}^{2\pi} F(\cos\theta, \sin\theta)d\theta = \int\limits_{|z|=1} F\left(\frac{z+z^{-1}}{2}, \frac{z-z^{-1}}{2i}\right)\frac{dz}{iz}$$

As a non-trivial example we evaluate

$$\int\limits_{0}^{2\pi} \frac{\sin^2\theta}{a+\cos\theta}d\theta \quad \text{for } a \in \mathbf{C}\backslash[-1,1]$$

where the restriction on a is chosen so that the denominator in the integrand does not vanish anywhere.

Using the above substitution we find

$$\int\limits_{0}^{2\pi} \frac{\sin^2\theta}{a+\cos\theta}d\theta = -\frac{1}{2i} \int\limits_{|z|=1} \frac{\left(z^2-1\right)^2}{z^2(z^2+2az+1)}dz$$

where the integrand, i.e. the function

$$f(z) = \frac{\left(z^2-1\right)^2}{z^2(z^2+2az+1)}$$

has a pole of order 2 at $z = 0$ and simple poles at

$$z = -a + \sqrt{a^2-1} \quad \text{and} \quad z = -a - \sqrt{a^2-1}$$

with corresponding residues $-2a$, $2\sqrt{a^2-1}$ and $-2\sqrt{a^2-1}$, where we to avoid ambiguity agree to take $\sqrt{a^2-1}$ as that branch of the square root of $a^2 - 1$ which is positive for $a > 1$. (Cf Example III.22(c)). To apply the Residue Theorem none of the poles

$$h(a) := -a - \sqrt{a^2-1} \quad \text{and} \quad g(a) := -a + \sqrt{a^2-1}$$

may belong to the unit circle $|z| = 1$. Since $h(a)$ and $g(a)$ are roots of the equation $z^2 + 2az + 1 = 0$, we have $h(a)g(a) = 1$ and $h(a) + g(a) = -2a$. Now if, say, $|h(a)| = 1$, then $g(a)$ is the complex conjugate of $h(a)$, so

$$\Re(h(a)) = \frac{h(a) + \overline{h(a)}}{2} = \frac{h(a) + g(a)}{2} = -a$$

But this means that a is real and $\sqrt{a^2-1}$ is purely imaginary, and that happens only in the excluded cases $a \in [-1,1]$. Since $|h(a)|$ never takes the value 1 and is > 1

for $a > 1$ we see that $|h(a)| > 1$ for all a considered, and so $|g(a)| < 1$ for all $a \in \mathbf{C}\backslash[-1,1]$. By the Residue Theorem

$$\int\limits_0^{2\pi} \frac{\sin^2\theta}{a+\cos\theta}d\theta = -\frac{1}{2i}2\pi i\{Res(f;0)+Res(f;g(a))\}$$

$$= -\pi\left\{-2a+2\sqrt{a^2-1}\right\} = 2\pi\left\{a-\sqrt{a^2-1}\right\}$$

The method of this example can be applied to any integral of the form

$$\int\limits_0^{2\pi} \frac{P(\cos t,\sin t)}{Q(\cos t,\sin t)}dt$$

where P and Q are polynomials in two variables and $Q(\cos t,\sin t) \neq 0$ for all $t \in [0,2\pi]$.

Example B.

In this example we want to evaluate integrals of the form $\int\limits_{-\infty}^{\infty} R(x)dx$ where R is a rational function. We state our result as a proposition:

Proposition 8.

Let P and Q be polynomials in one variable with $\deg P < \deg Q$ and assume that Q has no zeros on the real line.

(i) If $\mu > 0$ then

$$\lim_{R\to\infty}\int\limits_{-R}^{R} \frac{P(x)}{Q(x)}\,e^{i\mu x}dx$$

exists, and

$$\lim_{R\to\infty}\int\limits_{-R}^{R} \frac{P(x)}{Q(x)}\,e^{i\mu x}dx = 2\pi i\sum_w Res\left(\frac{P(z)}{Q(z)}\,e^{i\mu z};w\right)$$

where the sum ranges over $\{w \in \mathbf{C}\,|\,\Im w > 0\,,\,Q(w)=0\}$.

(ii) If $\deg(P) < \deg(Q) - 1$, then

$$\int\limits_{-\infty}^{\infty} \frac{P(x)}{Q(x)}dx$$

converges absolutely, and

$$\int\limits_{-\infty}^{\infty} \frac{P(x)}{Q(x)}dx = 2\pi i\sum_w Res\left(\frac{P}{Q};w\right)$$

where the sum ranges over $\{w \in \mathbf{C} \,|\, \Im w > 0 \,,\, Q(w) = 0\}$.

Proof:

(i) Note that there exist two positive constants A and B such that

$$\left|\frac{P(z)}{Q(z)}\right| \leq \frac{A}{|z|} \quad \text{when } |z| \geq B$$

Choose $R > B\sqrt{2}$ so large that the triangle in the picture below contains all the zeros of Q in the upper half plane.

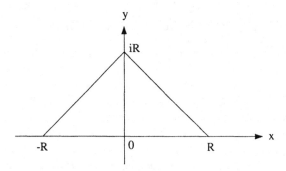

The distance from the origin to each of the two skew sides of the triangle is $R/\sqrt{2}$, so

$$\left|\frac{P(z)}{Q(z)}\right| \leq \frac{A\sqrt{2}}{R} \quad \text{for } z \in [R, iR]^* \cup [iR, -R]^*$$

With the usual parametrization of the segment $\gamma := [R, iR]$, i.e.

$$\gamma(t) = [R, iR](t) = R(it + 1 - t) \quad \text{for } 0 \leq t \leq 1$$

we find

$$\left|\int_{[R,iR]} \frac{P(z)}{Q(z)} e^{i\mu z} dz\right| = \left|\int_0^1 \frac{P(\gamma(t))}{Q(\gamma(t))} e^{i\mu R((i-1)t+1)} R(i-1) dt\right|$$

$$\leq \frac{A\sqrt{2}}{R} \int_0^1 e^{-\mu R t} R\sqrt{2}\, dt = \frac{2A}{\mu R}\left(1 - e^{-\mu R}\right) \leq \frac{2A}{\mu R} \to 0 \ \text{ as } R \to \infty$$

In exactly the same way we see that the integral along the other skew side $[iR, -R]$ of the triangle tends to zero as $R \to \infty$. The result (i) is then an immediate consequence of the Residue Theorem.

(ii) Proceed in the same way except for noting the stronger estimate

$$\left|\frac{P(z)}{Q(z)}\right| \leq \frac{A}{|z|^2} \text{ for } |z| \text{ large.}$$

□

Example C.

As an example of an integral involving a multivalued function we consider

$$\int\limits_0^\infty \frac{P(x)}{x^\alpha Q(x)} dx$$

We assume that $0 < \alpha < 1$ and that P and Q are polynomials in one variable such that $deg\,P < deg\,Q$ and such that Q has no zeros on $[0, \infty[$. Of course, $x^\alpha = \exp(\alpha\,Log\,x)$ in the integral.

It turns up to be a good idea to introduce that branch of the logarithm which is defined in $\Omega := \mathbf{C}\backslash[0, \infty[$ and has its imaginary part in the open interval $]0,2\pi[$. So for the remainder of this example, if $z \in \Omega$ we let

$$\log z = Log|z| + i\theta(z) \text{ , where } 0 < \theta(z) < 2\pi \text{ , and}$$
$$z^{-\alpha} = \exp(-\alpha \log z) = \exp(-\alpha\,Log|z| - i\alpha\theta(z))$$

We let f denote the function

$$f(z) := z^{-\alpha}\frac{P(z)}{Q(z)} \text{ for } z \in \Omega$$

It is holomorphic in Ω except at the zeros of Q.

Consider the following contour in which $R > 0$ is chosen so large, and r and ϵ, where $r > \epsilon > 0$, are chosen so small that the area inside contains all the zeros of Q. $\gamma_{\pm\epsilon}$ denote the two horizontal paths

$$\gamma_{\pm\epsilon}(t) = t \pm i\epsilon \text{ for } r - \epsilon \leq t \leq R - \epsilon$$

in the picture:

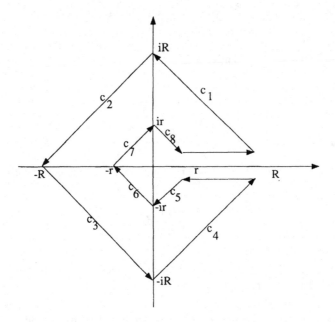

Then

$$\int\limits_{\gamma_\epsilon} f\,dz = \int\limits_{r-\epsilon}^{R-\epsilon} (t+i\epsilon)^{-\alpha} \frac{P(t+i\epsilon)}{Q(t+i\epsilon)}\,dt = \int\limits_{r-\epsilon}^{R-\epsilon} e^{-\alpha\,Log(t+i\epsilon)}\frac{P(t+i\epsilon)}{Q(t+i\epsilon)}\,dt$$

and similarly

$$\int\limits_{\gamma_{-\epsilon}} f\,dz = e^{-2\pi i\alpha} \int\limits_{r-\epsilon}^{R-\epsilon} e^{-\alpha\,Log(t-i\epsilon)}\frac{P(t-i\epsilon)}{Q(t-i\epsilon)}\,dt$$

so that

$$(*)\quad \lim_{\epsilon\to 0_+}\left\{ \int\limits_{\gamma_\epsilon} f\,dz - \int\limits_{\gamma_{-\epsilon}} f\,dz \right\} = \left(1-e^{-2\pi i\alpha}\right)\int\limits_{r}^{R} x^{-\alpha}\frac{P(x)}{Q(x)}\,dx$$

Let $c_1, c_2, \cdots, c_8$ denote the paths indicated in the figure. Using the abbreviation

$$\Sigma := \sum_{z\in\Omega} Res(f;z)$$

we get from the Residue Theorem that

$$(**)\quad \left| \int\limits_{\gamma_\epsilon} f\,dz - \int\limits_{\gamma_{-\epsilon}} f\,dz - \Sigma \right| \le \sum_{j=1}^{8} \left| \int\limits_{c_j} f\,dz \right|$$

so we shall estimate the eight integrals

$$\int_{c_j} f \, dz \quad , \quad j = 1, 2, \cdots, 8$$

Let us first consider the paths c_1, c_2, c_3, c_4 on the boundary of the large square. Here we infer from the assumption $\deg P < \deg Q$ that there exist constants $A > 0$, $B > 0$ such that

$$\left| \frac{P(z)}{Q(z)} \right| \le \frac{A}{|z|} \quad \text{for all } |z| \ge B$$

Noting that the distance from the origin to the boundary of the large square is $R/\sqrt{2}$ we get [choosing $R/\sqrt{2} \ge B$] for any outer path that

$$\left| \int_{c_j} f \, dz \right| \le \frac{1}{\left(R/\sqrt{2} \right)^\alpha} \frac{A}{R\sqrt{2}} R\sqrt{2} = \frac{A \, 2^{1 + \frac{\alpha}{2}}}{R^\alpha}$$

We shall next consider the boundary of the small square. Here we estimate P/Q by a constant $C > 0$ and find for any inner path c_j that

$$\left| \int_{c_j} f \, dz \right| \le \frac{C}{\left(r/\sqrt{2} \right)^\alpha} r\sqrt{2} \le C \, 2^{\frac{1+\alpha}{2}} r^{1-\alpha}$$

Substituting these estimates into (**) we get that

$$\left| \int_{\gamma_\epsilon} f \, dz - \int_{\gamma_{-\epsilon}} f \, dz \right| \le 4 \left[A \, 2^{1 + \frac{\alpha}{2}} R^{-A} + C \, 2^{\frac{1+\alpha}{2}} r^{1-\alpha} \right]$$
$$= C_1 R^{-\alpha} + C_2 r^{1-\alpha}$$

for some constants C_1 and C_2. Letting $\epsilon \to 0_+$ we read from (*) that

$$\left| \left(1 - e^{-2\pi i \alpha} \right) \int_r^R x^{-\alpha} \frac{P(x)}{Q(x)} dx - \Sigma \right| \le C_1 R^{-\alpha} + C_2 r^{1-\alpha}$$

Finally, letting $r \to 0$ and $R \to \infty$ we see that the right hand side converges to zero, so

$$\left(1 - e^{-2\pi i \alpha} \right) \int_0^\infty x^{-\alpha} \frac{P(x)}{Q(x)} dx - \Sigma = 0$$

The *final result* is now

$$\int_0^\infty x^{-\alpha} \frac{P(x)}{Q(x)} dx = \frac{\pi e^{\pi i \alpha}}{\sin(\pi\alpha)} \sum_{z \in \Omega} Res(f; z)$$

Section 4 Exercises

1. Let $z = 0$ be an isolated singularity of a function f, and

$$f(z) = \sum_{n=-\infty}^{\infty} a_n z^n$$

its Laurent expansion. Show that

(α) 0 is a removable singularity of f iff $a_n = 0$ for all $n < 0$.

(β) 0 is a pole of order m of f iff $a_{-m} \neq 0$ and $a_n = 0$ for all $n < -m$.

(γ) 0 is an essential singularity of f iff $a_n \neq 0$ for infinitely many negative n.

2. Evaluate

$$\int_0^{2\pi} e^{(e^{-i\theta})} d\theta$$

3. Find the value of

$$\frac{1}{2\pi i} \int_{|z|=1} \sin\left(\frac{1}{z}\right) dz$$

4. Let N be the year you were born. Find the value of

$$\frac{i}{4} \int_{|z|=N} \tan(\pi z) dz$$

5. Show that

$$\int_0^{2\pi} e^{\cos\theta} \cos(n\theta - \sin\theta) d\theta = \frac{2\pi}{n!} \text{ for } n = 0, 1, \ldots$$

6. Let P be a polynomial of degree ≥ 2. Show that

$$\int_{|z|=R} \frac{dz}{P(z)} = 0,$$

when R is chosen so large that the circle $|z| = R$ surrounds all the zeros of P.

7. Let P be a polynomial of degree $n \geq 2$ with distinct zeros $z_1, z_2, \cdots, z_n$. Show that

$$\sum_{j=1}^{n} \frac{1}{P'(z_j)} = 0$$

8. Let f be holomorphic in a neighborhood of $z = a$, and assume that $f(a) = 0$ and $f'(a) \neq 0$. Compute for $\epsilon > 0$ small the integral

$$\int_{|z-a|=\epsilon} z\frac{f'(z)}{f(z)} dz$$

9. Let m and n be integers ≥ 0. Find

$$\frac{1}{2\pi i} \int_{|z|=1} \left(z - a + \frac{1}{z-a} \right)^m (z-a)^{n-1} dz \quad \text{when } |a| \neq 1$$

Deduce *Wallis' formula*

$$\frac{1}{2\pi} \int_0^{2\pi} (2 \cos \theta)^{2m} d\theta = \frac{(2m)!}{(m!)^2}$$

10. Show that

$$\int_0^{2\pi} \frac{d\theta}{1 - 2r \cos \theta + r^2} = \frac{2\pi}{|1 - r^2|} \quad \text{when } r \in \mathbf{R} \backslash \{1, -1\}$$

11. Show that

$$\int_0^{2\pi} \frac{d\theta}{1 + a \cos \theta} = \frac{2\pi}{\sqrt{1-a^2}} \quad \text{for } 0 < a < 1$$

12. Show that

$$\int_0^{2\pi} \frac{d\theta}{1 + \mu^2 \cos^2 \theta} = \frac{2\pi}{\sqrt{1+\mu^2}} \quad \text{for } \mu \in \mathbf{R}$$

and derive the formula

$$\int_0^{2\pi} \frac{d\theta}{a^2 \cos^2 \theta + b^2 \sin^2 \theta} = \frac{2\pi}{ab} \quad \text{for } a > 0 \, , \, b > 0$$

13. Show that

$$\int_0^{\pi} \frac{d\theta}{(a + \sin^2 \theta)^2} = \pi \left(a + \frac{1}{2} \right) (a^2 + a)^{-\frac{3}{2}} \quad \text{when } a > 0$$

14. Show that

$$\int_{-\infty}^{\infty} \frac{e^{iax}}{1 + x^2} dx = \pi e^{-|a|} \quad \text{for } a \in \mathbf{R}$$

15. Show that

$$\int_0^{\pi} \tan (\theta + ib) d\theta = \pi i \, sign(b) \quad \text{for } b \in \mathbf{R} \backslash \{0\}$$

93

16. Pinpoint the error in the following reasoning : Let γ_R be the half circle in the upper half plane with center 0 and radius $R > 0$. Since the function (called Sinus Cardinalis)

$$Sinc(z) := \frac{\sin z}{z}$$

is entire (i.e. holomorphic in all of **C**), it follows from the Cauchy integral theorem with $f(z) = (Sinc\, z)^2$ that

$$\int\limits_{-\infty}^{\infty} f(x)dx = \lim_{R\to\infty}\int\limits_{-R}^{R} f(x)dx = -\lim_{R\to\infty}\int\limits_{\gamma_R} f(z)dz$$

Estimating sinus by 1 we get

$$\left|\int\limits_{\gamma_R}(Sinc\, z)^2 dz\right| \leq \int\limits_{0}^{\pi} R^{-2}Rd\theta = \frac{\pi}{R}\underset{R\to\infty}{\to} 0$$

so that

$$\int\limits_{-\infty}^{\infty}(Sinc\, x)^2 dx = 0$$

Show that the correct value of the integral isn't 0, but π.

17. Show for $n = 1, 2, \cdots$ that

$$\int\limits_{-\infty}^{\infty}(Sinc\, x)^n dx = \frac{\pi}{2^{n-1}(n-1)!}\sum_{k<\frac{n}{2}}\binom{n}{k}(-1)^k(n-2k)^{n-1}$$

18. Show that

$$\int\limits_{-\infty}^{\infty}\frac{x^3\sin x}{(x^2+1)^2}dx = \frac{\pi}{2e}$$

19. Show that

$$\int\limits_{0}^{\infty}\frac{x\sin x}{x^4+1}dx = \frac{\pi}{2}e^{-\frac{1}{\sqrt{2}}}\sin\left(\frac{1}{\sqrt{2}}\right)$$

20. Show that

$$\int\limits_{-\infty}^{\infty}\frac{\cos x}{\cosh x}dx = \frac{\pi}{\cosh\left(\frac{\pi}{2}\right)}$$

Hint: Consider the contour of height π

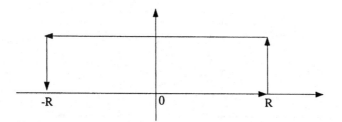

21. Show for $n = 2, 3, \cdots$ that

$$\int_0^\infty \frac{dx}{1+x^n} = \frac{\frac{\pi}{n}}{\sin\left(\frac{\pi}{n}\right)}$$

by help of the following contour in which $P = R\exp\left(2\pi i/n\right)$

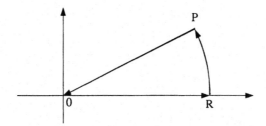

Compute the more general expression

$$\int_0^\infty \frac{x^\alpha}{1+x^n} dx \quad \text{where} \quad -1 < \alpha < n-1$$

Answer :

$$\frac{\pi/n}{\sin\left((1+\alpha)\pi/n\right)}$$

22. Show that

$$\int_0^\infty \frac{Log\, x}{\left(x^2+1\right)^2} dx = -\frac{\pi}{4}$$

Hint : Consider the contour

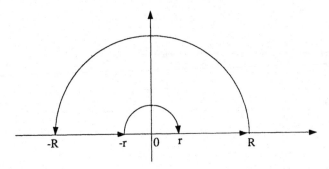

An elementary deduction of the result can be found in [AGR].

23. Show that

$$\int\limits_0^\infty \frac{Log\,x}{x^2-1}dx = \frac{\pi^2}{4}$$

Hint: Consider the contour

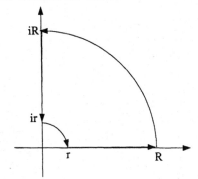

24. In this problem we will deduce the following

Theorem.

If f is a rational function such that $|f(z)| = 1$ whenever $|z| = 1$, then f has off its poles the form

$$f(z) = cz^m \prod_{k=1}^n \frac{z - a_k}{1 - \overline{a_k}z}$$

where c is a constant of modulus 1, m is an integer, $n \in \{0, 1, 2, \cdots\}$, and where $a_1, a_2, \cdots, a_n$ are complex constants such that $|a_k| \neq 1$ for all k.

Proceed as follows for the proof:

a. Show that we in the proof without loss of generality may assume that 0 is neither a zero nor a pole of f [divide by an appropriate power of z].

96

b. Let $a \in \mathbb{C}$. Show that

$$\left| \frac{z-a}{1-\bar{a}z} \right| = 1 \text{ for all } z \in S^1 \backslash \{a\}$$

c. Let $z_1, z_2, \cdots, z_r$ denote the zeros of f inside the unit disc, and $p_1, p_2, \cdots, p_s$ its poles there, zeros as well as poles repeated according to their multiplicity. Show that the function

$$F(z) := f(z) \prod_{k=1}^{r} \frac{1 - \overline{z_k}z}{z - z_k} \prod_{l=1}^{s} \frac{z - p_l}{1 - \overline{p_l}z}$$

is holomorphic and zero-free in a neighborhood of $B[0,1]$ and that $|F(z)| = 1$ whenever $|z| = 1$.

d. Show that F is a constant in $B(0,1)$ and hence everywhere.

e. Finish the proof. $\qquad\qquad\square$

Chapter 7 The Picard Theorems

In this chapter we shall prove Picard's "big" theorem on the strange behavior of holomorphic functions near essential singularities, and as a consequence derive his "little" theorem on entire functions. These two theorems have been the spur for much development in complex analysis, in particular for the so-called Nevanlinna theory which is concerned with the distribution of the values of meromorphic functions. Let us cite the two theorems:

Theorem 1 (Picard's little theorem).
The range of a nonconstant entire function omits at most one complex number.

Theorem 2 (Picard's big theorem).
Let f be holomorphic in $\Omega \backslash \{z_0\}$, where Ω is an open subset of the complex plane and $z_0 \in \Omega$.

If f has an essential singularity at z_0 then $f(\Omega \backslash \{z_0\})$ is either $\mathbf{C}$ or $\mathbf{C}$ with one point removed.

Section 1 Liouville's and Casorati-Weierstrass' theorems

Liouville's theorem (Theorem I.11) gives a first weak hint to the truth of Picard's little theorem, so we repeat it here:

Theorem 3 (Liouville's theorem).
Let f be entire.
(i) If f is bounded then f is constant.
(ii) More generally, if f for some real constants a, b and m satisfies the estimate

$$|f(z)| \leq a|z|^m \ \text{for all } |z| > b$$

then f is a polynomial of degree $\leq m$.

As an important application of Liouville's theorem we derive the fundamental theorem of algebra which asserts that every algebraic equation over the field of complex numbers has a root. More generally that it takes any complex value and thus shows the validity of Picard's little theorem for polynomials. The fundamental theorem of algebra is important because it implies that any polynomial can be written as a product of linear factors.

Corollary 4 (The fundamental theorem of algebra).
Let $a_0, a_1, \cdots, a_n \in \mathbf{C}$ where $n \geq 1$ and $a_n \neq 0$, and let p denote the polynomial

$$p(z) := a_0 + a_1 z + \cdots + a_n z^n \ \text{for } z \in \mathbf{C}$$

Then there exists a $z_0 \in \mathbf{C}$ such that $p(z_0) = 0$.

Although the fundamental theorem of algebra looks like an algebraic result, being concerned with polynomials, it also involves the completeness of the reals: It is obviously false if $\mathbf{C} = \mathbf{R} + i\mathbf{R}$ is replaced by $\mathbf{Q} + i\mathbf{Q}$, where $\mathbf{Q}$ denotes the rational numbers. For a discussion and many references see p.117-118 of the monography [Bu].

Proof: The proof goes by contradiction, so we assume that p never vanishes. Then the function $f(z) := 1/p(z)$ is entire and converges to 0 as $|z| \to \infty$. In particular f is bounded and so a constant by Liouville's theorem. But then p is constant, contradicting that the degree of p is ≥ 1. $\qquad\qquad\qquad\qquad\qquad\qquad\qquad\qquad\Box$

The Casorati-Weierstrass' theorem is an immediate consequence of Picard's big theorem. We give a direct proof though, since it is quite simple. In the Russian literature the theorem is attributed to Y.V. Sokhotskii. For a detailed history of it see [Neu].

Theorem 5 (Casorati-Weierstrass' theorem).
Let f have an essential singularity at $z_0 \in \mathbf{C}$. Then for any $w \in \mathbf{C}$ there exists a sequence $z_1, z_2, \cdots$ such that $z_n \to z_0$ and $f(z_n) \to w$ as $n \to \infty$.

Proof: The proof goes by contradiction : If the assertion were false then there would exist a $w_0 \in \mathbf{C}$ and an $\epsilon > 0$ and a neighborhood Ω of z_0 such that

$$|f(z) - w_0| \geq \epsilon \ \text{ for all } \ z \in \Omega \backslash \{z_0\}$$

The function

$$g(z) := \frac{1}{f(z) - w_0} \ , \ z \in \Omega \backslash \{z_0\}$$

is then analytic in $\Omega \backslash \{z_0\}$ and bounded (by $1/\epsilon$), so z_0 is a removable singularity for g. Thus

$$f(z) = w_0 + \frac{1}{g(z)}$$

is either analytic at $z = z_0$ (if $g(z_0) \neq 0$), or has a pole at $z = z_0$ (if $g(z_0) = 0$). In any case it contradicts our assumption about z_0 being an essential singularity. $\qquad\Box$

Section 2 Picard's two theorems

Our key to the proof of Picard's big theorem is the following remarkable result on the range of functions which are analytic in the unit disc.

Theorem 6 (Bloch-Landau's theorem).

There exists a positive number A with following property: Let f be any function, holomorphic in the unit disc and such that $|f'(0)| \geq 1$. Then the range of f contains a disc of radius A.

The surprising feature of the theorem is of course the existence of the universal constant A in spite of the vast class of functions involved. For our purposes any positive constant will suffice.

Remark.

Let us for

$$f \in \Phi := \left\{ g \in Hol(B(0,1)) \big| |g'(0)| \geq 1 \right\}$$

put

$$L(f) := \sup \left\{ r > 0 \mid f(B(0,1)) \text{ contains a disc of radius } r \right\}$$

The theorem above then tells us that *Landau's constant*

$$L := \inf_{f \in \Phi} \{L(f)\}$$

is positive. Our proof of Theorem 6 reveals that $L \geq \frac{1}{13}$. The exact value of Landau's constant is not known, but it has been ascertained that $0.5 \leq L \leq 0.56$.

Theorem 6 is an immediate corollary of an even more surprising quantitative discovery on the range of an analytic function :

Theorem 7 (Bloch's theorem).

Let f be any function which is holomorphic in $B(0,1)$ and has $|f'(0)| \geq 1$.

Then there exists a subdisc D_f of $B(0,1)$ such that $f(D_f)$ contains a disc of radius 0.43 and f is univalent on D_f.

Remark .

Let $B(f)$ be the supremum of all $r > 0$ for which there exists a region D on which f is univalent and such that $f(D)$ contains a disc of radius r. Then Bloch's theorem tells us that *Bloch's constant*

$$B := \inf \left\{ B(f) \mid f \in Hol(B(0,1)) \text{ and } |f'(0)| \geq 1 \right\}$$

is bigger than 0.43. The exact value of Bloch's constant is not known, although it is known that $0.43 \leq B \leq 0.47$.

For interesting gossip on André Bloch's life see the essay [Ca] and the follow up [CF].

Proof of Theorem 6:

We may assume that $|f'(0)| = 1$. We will first treat the special case where f is holomorphic on a neighborhood of the closed disc $B[0,1]$.

The function $a : [0,1] \to [0, \infty[$ given by

$$a(r) := (1 - r) \sup_{|z|=r} \{|f'(z)|\} \quad \text{for } 0 \le r \le 1$$

is continuous. Since $a(0) = 1$ and $a(1) = 0$, there exists a largest number $s \in [0, 1[$ such that $a(s) = 1$. Choose $\zeta \in B(0,1)$ such that $|\zeta| = s$ and $|f'(\zeta)| = \sup_{|z|=s} \{|f'(z)|\}$.

Consider for $R := (1 - s)/2$ the function

$$F(z) := 2(f(Rz + \zeta) - f(\zeta)) \quad \text{for } |z| \le 1$$

F is holomorphic on $B(0,1)$, $F(0) = 0$, and

$$|F'(0)| = 2R|f'(\zeta)| = \frac{2Ra(s)}{1 - s} = 1$$

Furthermore, since $a(r) < 1$ when $s < r$, we get

$$\begin{aligned}
\frac{|F'(z)|}{2} &= R|f'(Rz + \zeta)| \le R \sup \{|f'(w)| \,|\, |w| \le s + R\} \\
&= R \sup \{|f'(w)| \,|\, |w| = s + R\} \\
&= \frac{R}{1 - s - R} a(s + R) < \frac{R}{1 - s - R} = 1
\end{aligned}$$

so $|F'(z)| \le 2$.

In the following lemma we shall show that the range of F contains a disc of radius $1/6$. From the definition of F we see that the range of f then contains a disc of radius $1/12$. This takes care of the special case.

In the general case we consider for any $0 < \rho < 1$ the function $z \to f(\rho z)/\rho$. It satisfies the conditions of the special case, so its range contains a disc of radius $1/12$. But that means that the range of f contains a disc of radius $\rho/12$. Taking $\rho = 12/13$ we see that the range of f contains a disc of radius $1/13$. $\qquad\square$

Lemma 8.

Let $F \in Hol(B(0,1))$ satisfy that $F(0) = 0$, $F'(0) = 1$ and $|F'| \le M$, where M is a constant. Then

$$F(B(0,1)) \supseteq B\left(0, \frac{1}{2(M + 1)}\right)$$

Proof: The function

$$z \to \frac{F'(z) - 1}{M + 1}$$

satisfies the conditions of Schwarz' lemma, so

$$|F'(z) - 1| \le (M+1)|z| \quad \text{for all } |z| < 1$$

Integrating $F'(z) - 1$ along the segment $[0, z]$ we get the inequality

$$|F(z) - z| \le \frac{M+1}{2}|z|^2$$

It says in particular that we for ζ on the circle $|\zeta| = 1/(M+1)$ have $|F(\zeta) - \zeta| \le \frac{1}{2(M+1)}$. If furthermore $w \in B\left(0, \frac{1}{2(M+1)}\right)$ then

$$|(F(\zeta) - w) - (\zeta - w)| = |F(\zeta) - \zeta| \le \frac{1}{2(M+1)} < |\zeta - w|$$

so Rouché's theorem (Theorem IV.13) tells us that the function $z \to F(z) - w$ has a zero in the ball $B(0, 1/(M+1))$. Since w was arbitrary we have shown that

$$F\left(B\left(0, \frac{1}{M+1}\right)\right) \supseteq B\left(0, \frac{1}{2(M+1)}\right)$$

which is more than required. $\qquad\qquad\qquad\qquad\qquad\qquad\qquad\qquad\qquad$ $\square$

We need one result more prior to Picard's big theorem, viz. Schottky's theorem.

Theorem 9 (Schottky's theorem).

Let $M > 0$ and $r \in]0, 1[$ be given. Then there exists a constant $C > 0$ such that the following implication holds :

If F is holomorphic in $B(0, 1)$, omits 0 and 1 from its range, and if $|F(0)| \le M$, then $|F(z)| \le C$ for all $z \in B(0, r)$.

Proof: Since F does not assume the value 0 in the starshaped set $B(0, 1)$, F has a holomorphic logarithm $\log F$ [Proposition III.10] , which we choose such that $|\Im(\log F(0))| \le \pi$. The function $A := (2\pi i)^{-1} \log F$ does not assume integer values, because F never takes the value 1. Let $\sqrt{A}$ and $\sqrt{A-1}$ be holomorphic square roots of A and $A - 1$ (They exist by Proposition III.21). Then $B := \sqrt{A} - \sqrt{A-1}$ is holomorphic in $B(0, 1)$, vanishes nowhere and cannot assume the values $\sqrt{n} \pm \sqrt{n-1}$ for $n = 1, 2, \cdots$. [Indeed, if

$$\sqrt{A(z)} - \sqrt{A(z) - 1} = \sqrt{n} \pm \sqrt{n-1} \quad \text{for some } z \text{ and } n$$

then taking reciprocals we get

$$\sqrt{A(z)} + \sqrt{A(z) - 1} = \sqrt{n} - \sqrt{n-1} \text{ resp. } \sqrt{n} + \sqrt{n-1}$$

Adding and squaring shows that $A(z) = n$, which is excluded.].

Since B never vanishes there is a holomorphic branch H of $\log B$, and H cannot assume the values

$$a_{n,m} := Log\left\{\sqrt{n} \pm \sqrt{n-1}\right\} + 2\pi i m \quad \text{for } n = 1, 2, \cdots \text{ and } m \in \mathbf{Z}$$

We leave it to the reader to verify that every disc of radius 10 contains at least one of the points $a_{n,m}$, so that the range of H cannot cover any disc of radius 10.

If $z \in B(0,1)$ and $H'(z) \neq 0$, then the values of the function

$$\zeta \to \frac{H(\zeta) - H(z)}{H'(z)} \quad \text{for } \zeta \in B(z, 1 - |z|)$$

must fill a disc of radius $(1 - |z|)/13$ (by our proof of the Bloch-Landau theorem), so the values of H fill a disc of radius $|H'(z)|(1 - |z|)/13$. This quantity cannot exceed 10, so

$$(1) \quad \left|H'(z)\right|(1 - |z|) \leq 130$$

Although (1) was derived under the assumption $H'(z) \neq 0$, it is clearly also valid when $H'(z) = 0$. Since $H(z) - H(0)$ is the integral of H' along the segment $[0, z]$, the estimate (1) leads to

$$(2) \quad |H(z)| \leq |H(0)| + 130\,Log\,(1 - |z|)^{-1} \leq |H(0)| - 130\,log\,(1 - r)$$
$$\text{for } |z| \leq r$$

Now recall the definition of H :

$$exp\,H = \sqrt{\frac{\log F}{2\pi i}} - \sqrt{\frac{\log F}{2\pi i} - 1}$$

Taking reciprocals and adding the result we get

$$(3) \quad \sqrt{\frac{\log F(z)}{2\pi i}} = \frac{e^{H(z)} + e^{-H(z)}}{2}$$

so that

$$F = -exp\left\{\frac{\pi i}{2}\left(e^{2H} + e^{-2H}\right)\right\} \quad \text{and} \quad |F| \leq exp\left\{\pi e^{2|H|}\right\}$$

In view of (2) the theorem follows once we establish that $|H(0)| \leq C_1$ where C_1 is a constant depending only on the bound M on $F(0)$.

We will first treat those functions that satisfy the extra requirement $|F(0)| \geq \frac{1}{2}$. For such a function F we get from (3) that there is a constant C_2, depending only on M such that

$$C_2 \geq \left|\frac{e^{H(0)} + e^{-H(0)}}{2}\right| \geq \frac{e^{\Re(H(0))} - e^{-\Re(H(0))}}{2} = \sinh\left(\Re(H(0))\right)$$

which gives us an upper bound of the desired type on $\Re(H(0))$. Similarly for a lower bound on $\Re(H(0))$. The imaginary part poses no problems since we always may choose $H = \log B$ such that $|\Im(H(0))| \leq \pi$. We have now proved the theorem under the extra requirement that $|F(0)| \geq \frac{1}{2}$.

If $|F(0)| < \frac{1}{2}$ then we apply the just derived result to the function $1 - F$ instead of F. $\qquad\qquad\qquad\qquad\qquad\qquad\qquad\qquad\qquad\qquad\qquad\qquad\qquad\quad$ $\square$

We finally arrive at Picard's big theorem, which is a remarkable generalization of the Casorati-Weierstrass theorem.

Theorem 10 (Picard's big theorem).

If F has an essential singularity then the range of F omits at most one complex number.

The example $F(z) = \exp(1/z)$ shows that it is too much to hope that the range of F is all of $\mathbf{C}$.

Proof: By translation in $\mathbf{C}$ we may assume that the singularity is situated at $z = 0$, and by dilation that F is holomorphic in $B(0, e^{2\pi}) \setminus \{0\}$. We will show that if F omits two complex numbers, say a and b, then 0 is either a pole or a removable singularity; that will prove the theorem. We may even assume that F omits the special values 0 and 1 - if not we replace F by

$$z \rightarrow \frac{F(z) - a}{b - a}$$

Case 1: $|F(z)| \rightarrow \infty$ as $z \rightarrow 0$.

In this case 0 is a pole of F.

Case 2: There exists a sequence $z_n \underset{n \rightarrow \infty}{\rightarrow} 0$ such that $\{|F(z_n)|\}$ remains bounded, say $|F(z_n)| \leq M < \infty$ for all n.

Passing to a subsequence we may assume that

$$1 > |z_1| > \cdots > |z_n| > |z_{n+1}| > \cdots \underset{n \rightarrow \infty}{\rightarrow} 0$$

Consider for fixed n the function $\zeta \rightarrow F(z_n e^{2\pi i \zeta})$ which is holomorphic in $B(0, 1)$, omits the values 0 and 1 and has $|F(z_n e^{2\pi i 0})| \leq M$. By Schottky's theorem there exists a constant C, depending only on the bound M, such that

$$\left| F\left(z_n e^{2\pi i \zeta}\right) \right| \leq C \text{ for all } \zeta \in B\left[0, \frac{1}{2}\right]$$

In particular

$$\left| F\left(z_n e^{2\pi i t}\right) \right| \leq C \text{ for all } t \in \left[-\frac{1}{2}, \frac{1}{2}\right]$$

so that $|F|$ is bounded by C on the circle $\{z \in \mathbf{C} \mid |z| = |z_n|\}$.

Since the constant C is independent of n we get by the weak maximum principle that $|F| \leq C$ on all of $B(0, |z_1|)\backslash\{0\}$. But then 0 is a removable singularity of F. $\square$

The next result, Picard's little theorem, is an extension of the fundamental theorem of algebra and of Liouville's theorem.

Theorem 11 *(Picard's little theorem).*

If F is a non-constant entire function then the range of F omits at most one complex number.

The function $F(z) = \exp z$ demonstrates that it does happen that a complex number is absent from the range of F.

Proof: Consider the function

$$G(z) := F\left(\frac{1}{z}\right) \ \text{ for } \ z \in \mathbf{C} \backslash \{0\}$$

If 0 is an essential singularity for G then we are done by Picard's big theorem. If 0 is a pole of order m or a removable singularity of G (a pole of order $m = 0$) then G can be written in the form $G(z) = z^{-m}H(z)$ where H is holomorphic in all of $\mathbf{C}$, also at $z = 0$. Now $F(z) = z^m H\left(\frac{1}{z}\right)$ so that

$$|F(z)| \leq (|H(0)| + 1)|z|^m \ \text{ for } |z| \text{ large}$$

By Liouville's theorem F is a polynomial, and so its range includes all of $\mathbf{C}$ by the fundamental theorem of algebra. $\square$

Section 3 Exercises

1. Let F be an entire function with the property that

$$\frac{F(z)}{z} \to 0 \ \text{ as } \ |z| \to \infty$$

Show that F is a constant.

2. Let F be an entire function with the property that

$$|F(z)| \geq 1 \ \text{ for all } \ z \in \mathbf{C}$$

Show that F is a constant.

3. Let F be an entire function with the property that $\Re F$ is bounded from above. Show that F is a constant (Hint: Consider $\exp F$).

4. (Injectiveness of the Fourier transform. Cf. [New]).

Let f be continuous and absolutely integrable over $\mathbf{R}$ and assume that

$$\int_{-\infty}^{\infty} e^{-ix\xi} f(x)\,dx = 0 \quad \text{for all } \xi \in \mathbf{R}$$

(a) Show for each fixed $a \in \mathbf{R}$ that

$$\int_{-\infty}^{a} e^{-i\xi(x-a)} f(x)\,dx = -\int_{a}^{\infty} e^{-i\xi(x-a)} f(x)\,dx$$

and let $F_a(\xi)$ denote this common value.

(b) Show that the left hand side for fixed $a \in \mathbf{R}$ can be extended to a function which is continuous and bounded in the upper half plane $\Im\xi \geq 0$ and is holomorphic in the open half plane $\Im\xi > 0$.

Show similar statements for the right hand side.

(c) Show successively that F_a for fixed $a \in \mathbf{R}$ is holomorphic in all of $\mathbf{C}$, constant, even 0.

(d) Show that $f = 0$.

(e) Show by induction on n that the same result holds for functions f which are integrable over $\mathbf{R}^n$.

5. Let F be an entire function that never takes values in $]0, \infty[$. Show that F is a constant.

6. Let F be a rational function, i.e. a quotient between two polynomials. Assume that its poles occur at $a_1, a_2, \cdots, a_n$ with orders $m_1, m_2, \cdots, m_n$.

Show that there exist a polynomial p and constants $c_{jk} \in \mathbf{C}$ such that

$$F(z) = p(z) + \sum_{j=1}^{n} \sum_{k=1}^{m_j} \frac{c_{jk}}{(z - a_j)^k} \quad \text{for } z \in \mathbf{C}\backslash\{a_1, a_2, \cdots, a_n\}$$

7. Let F be holomorphic in $\mathbf{C}$ with the exception of finitely many poles. Assume that $F(1/z)$ has an inessential singularity at $z = 0$. Show that F is a rational function, i.e. a quotient between two polynomials.

8. Show the following version of Picard's little theorem:

Let F be a non-polynomial, entire function. Then there exists a $z_0 \in \mathbf{C}$ such that F takes every value in $\mathbf{C}\backslash\{z_0\}$ infinitely often.

Hint: Assume the conclusion fails. Show by Picard's big theorem that the function $z \to F(\frac{1}{z})$ only has an inessential singularity at $z = 0$. Apply Liouville's theorem.

9. Let F be a non-constant, entire function. Assume that F does not take the value $a \in \mathbf{C}$. Let $b \in \mathbf{C}\backslash\{a\}$. Show that F takes the value b infinitely often.

10. Let F be entire and injective. Show that F is a polynomial of degree 1.

11. Define for fixed $\epsilon > 0$ the function $f : B(0,1) \to \mathbf{C}$ by

$$f(z) := \frac{\epsilon}{2}\left\{\left(\frac{1+z}{1-z}\right)^{\frac{1}{\epsilon}} - 1\right\} \quad \text{for } z \in B(0,1)$$

(α) Show that f has $f'(0) = 1$ and hence satisfies the conditions in Bloch-Landau's theorem.

(β) Show that $-\epsilon/2$ does not belong to the range of f.

(γ) Taking $\epsilon = 1/13$ we see that $f(B(0,1))$ does not contain a ball around 0 with radius 1/13. Doesn't that contradict Bloch-Landau's theorem?

12. Let S_0 denote the set of functions $f \in Hol(B(0,1))$ which satisfy $f(0) = 0$, $f'(0) = 1$, and $f(z) = 0$ only if $z = 0$.

We want to show that there exists a universal constant $\alpha > 0$ such that $f \in S_0$ implies that $B(0, \alpha) \subseteq f(B(0,1))$.

We prove it by contradiction. So assume that the implication is false.

(a) Show that there exist sequences $\{\alpha_n\} \subseteq \mathbf{C}\backslash\{0\}$ such that $\alpha_n \underset{n\to\infty}{\to} 0$ and $\{f_n\} \subseteq S_0$ such that $\alpha_n \notin f_n(B(0,1))).$

(b) Show that

$$1 - \frac{f_n}{\alpha_n} \; : \; B(0,1) \to \mathbf{C}\backslash\{0\}$$

has a holomorphic logarithm L_n such that $L_n(0) = 0$.

(c) Show that the sequence $\{L_n\}$ is uniformly bounded on B(0,1/2) .

(d) Show that $\{L_n'(0)\}$ is a bounded sequence.

(e) Derive a contradiction !

Section 4 Alternative treatment

The following alternative treatment of Schottky's and Picard's theorems is borrowed from the paper [MS].

$$\Delta = \frac{\partial^2}{\partial x^2} + \frac{\partial^2}{\partial y^2}$$

denotes the *Laplace operator* = the *Laplacian* on $\mathbf{R}^2$. In polar coordinates (r, θ) it takes the form

$$\Delta = \frac{\partial^2}{\partial r^2} + \frac{1}{r}\frac{\partial}{\partial r} + \frac{1}{r^2}\frac{\partial^2}{\partial \theta^2}$$

We shall a couple of times use that $\Delta F = 0$ when F is a holomorphic function. That is a consequence of the Cauchy-Riemann equations (See Theorem XII.2 for more information on this point).

Definition 12.

Let Ω be an open subset of the complex plane. A continuous function $\rho : \Omega \to [0, \infty[$ is said to be a pseudo-metric *if it is of class C^2 on $\{z \in \Omega \,|\, \rho(z) > 0\}$. A* metric *is a function in $C^2(\Omega,]0, \infty[)$, so a metric is also a pseudo-metric.*

The (Gaussian) curvature *of a pseudo-metric* ρ *is the function* κ *defined by*

$$\kappa(z,\rho) := -\frac{\Delta(Log\,\rho(z))}{\rho(z)^2}$$

in the subset of Ω *where* $\rho(z) > 0$.

Examples 13.

(i)

$$\lambda_R(z) := \frac{2R}{R^2 - |z|^2}$$

is a metric on $B(0,R)$ with constant curvature -1.

(ii)

$$\sigma(z) := \frac{2}{1 + |z|^2}$$

is a metric on **C** with constant curvature $+1$.

The formula in the following lemma expresses that the curvature is a conformal invariant, i.e. is preserved under holomorphic mappings, and so it is a natural object that ought to be studied.

Lemma 14.

Let $f : \Omega \to \Omega'$ *be a holomorphic map between two open subsets* Ω *and* Ω' *of the complex plane, and let* ρ *be a metric on* Ω'. *Then* $f^*(\rho) := \rho \circ f\,|f'|$ *is a pseudo-metric on* Ω *and*

$$\kappa(z, f^*(\rho)) = \kappa(f(z),\rho) \text{ for all } z \in \Omega \text{ for which } f'(z) \neq 0$$

Proof: It is obvious that $f^*(\rho)$ is a pseudo-metric, so left is the formula for the curvature, i.e. that

$$-\frac{\Delta(Log(\rho \circ f\,|f'|))}{\rho \circ f\,|f'|^2}(z) = -\frac{(\Delta\,Log\,\rho)(f(z))}{\rho(f(z))^2}$$

or in other words that

$$\Delta\{Log(\rho \circ f) + Log\,|f'|\}(z) = \{\Delta\,Log\,\rho\}(f(z))\,|f'(z)|^2$$

Here the term $\Delta(Log\,|f'|)$ on the left hand side vanishes, because $Log\,|f'|$ (locally) is the real part of a holomorphic function (branches of $log\,f'$), so it is left to show the formula

$$\{\Delta(F \circ f)\}(z) = (\Delta F)(f(z))\,|f'(z)|^2$$

a formula that may be left to the reader. $\qquad\square$

Theorem 15 (Ahlfors' lemma).

λ_1 *is the biggest pseudometric on* $B(0,1)$ *among the ones with curvature at most* -1.

Proof: Let us for arbitrary but fixed $r \in]0,1[$ consider the function $v := \rho/\lambda_r$ on $B[0,r]$. It suffices to show that $v \leq 1$, i.e. that $\rho \leq \lambda_r$ because we then obtain the desired conclusion by letting $r \to 1$.

Since v is non-negative and $v(z) \to 0$ as $|z| \to r$, we see that the continuous function v attains its supremum m at some point z_0 in the open ball $B(0,r)$. We shall show that $m \leq 1$. If $m = 0$ we are done, so we may as well assume that $\rho(z_0) > 0$. Since z_0 is a maximum for v and hence for $\text{Log}\,v$, we get that $0 \geq (\Delta\,Log\,v)(z_0)$.
Now,

$$0 \geq (\Delta\,Log\,v)(z_0) = (\Delta\,Log\,\rho)(z_0) - (\Delta\,Log\,\lambda_r)(z_0)$$
$$= -\kappa(z_0,\rho)\rho(z_0)^2 + \kappa(z_0,\lambda_r)\lambda_r(z_0)^2$$
$$= -\kappa(z_0,\rho)\rho(z_0)^2 - \lambda_r(z_0)^2$$

By the assumption on the curvature of ρ, we find $0 \geq \rho(z_0)^2 - \lambda_r(z_0)^2$, so $m = \rho(z_0)/\lambda_r(z_0) \leq 1$. $\qquad\square$

Proposition 16.

There exists a metric ρ *on* $\mathbf{C}\setminus\{0,1\}$ *with the following two properties:*

(i) ρ *has curvature at most* -1.

(ii) $\rho \geq c\sigma$ *for some constant* $c > 0$.

Proof: Using the Laplacian in polar coordinates we find by brute force computations for any $\alpha \in \mathbf{R}$ and $z \in \mathbf{C}\setminus\{0\}$ that

$$\Delta\big(Log(1 + |z|^\alpha)\big) = \alpha^2\frac{|z|^{\alpha-2}}{\big(1 + |z|^\alpha\big)^2}$$

Furthermore $\Delta(Log\,|z|) = 0$ because $Log\,|z|$ locally is the real part of a holomorphic function (branches of $\log z$). Combining these two facts we get for any $\alpha,\beta \in \mathbf{R}$ and $z \in \mathbf{C}\setminus\{0\}$ that

$$\Delta\left\{Log\,\frac{1 + |z|^\alpha}{|z|^\beta}\right\} = \alpha^2\frac{|z|^{\alpha-2}}{\big(1 + |z|^\alpha\big)^2}$$

Since the Laplacian $\Delta = \partial^2/\partial x^2 + \partial^2/\partial y^2$ has constant coefficients it satisfies that

$$(\Delta F)(z - z_0) = \Delta(w \to F(w - z_0))(z)$$

so in particular for $z \in \mathbf{C}\setminus\{1\}$

$$\Delta\left\{Log\,\frac{1 + |z-1|^\alpha}{|z-1|^\beta}\right\} = \alpha^2\frac{|z-1|^{\alpha-2}}{\big(1 + |z-1|^\alpha\big)^2}$$

We claim that for suitable $\alpha > 0$, $\beta > 0$ a positive multiple of the metric

$$\tau(z) := \frac{1 + |z|^\alpha}{|z|^\beta} \frac{1 + |z - 1|^\alpha}{|z - 1|^\beta} \quad \text{for } z \in \mathbf{C} \backslash \{0, 1\}$$

can be used as the desired metric ρ.

By a straightforward calculation

$$\kappa(z, \tau) = - \alpha^2 \frac{|z|^{\alpha + 2\beta - 2}}{\left(1 + |z|^\alpha\right)^4} \frac{|z - 1|^{2\beta}}{\left(1 + |z - 1|^\alpha\right)^2}$$
$$- \alpha^2 \frac{|z - 1|^{\alpha + 2\beta - 2}}{\left(1 + |z - 1|^\alpha\right)^4} \frac{|z|^{2\beta}}{\left(1 + |z|^\alpha\right)^2}$$

so the curvature is everywhere negative.

Taking $\alpha = 1/6$ and $\beta = 3/4$ we find that

(1) $\displaystyle \lim_{|z| \to \infty} \frac{\tau(z)}{\sigma(z)} = +\infty$ and (2) $\lim K(z, \tau) = -\infty$ when $z \to 0, 1$ or ∞

Combining (2) with the fact that the curvature is negative we find that there exists a constant $k > 0$ such that $\kappa(z, \tau) \leq -k$ on $\mathbf{C} \backslash \{0, 1\}$. Then $\rho := \sqrt{k}\tau$ has curvature at most -1, which is the first property.

From (1) we see that

$$\lim_{|z| \to \infty} \frac{\rho(z)}{\sigma(z)} = +\infty$$

which implies the second property. $\square$

Theorem 17 (Schottky's theorem).

To each $M > 0$ and $r \in \,]0, 1[$ there exists a constant $C > 0$ such that the following implication holds :

If $f \in Hol(B(0, 1))$, $|f(0)| \leq M$ and the range of f omits 0 and 1 , then $\sup_{|z| \leq r} \{|f(z)|\} \leq C$.

Proof: Let ρ and c be as in Proposition 16. Then $f^*(\rho)$ is a pseudo-metric on B(0,1) with curvature at most -1 (Lemma 14), by Ahlfors' lemma $f^*(\rho) \leq \lambda_1$, and so $f^*(\sigma) \leq c^{-1}f^*(\rho) \leq c^{-1}\lambda_1$, i.e.

$$\frac{|f'(z)|}{1 + |f(z)|^2} \leq \frac{c^{-1}}{1 - |z|^2} \quad \text{for all } z \in B(0, 1)$$

and so

$$\frac{|f'(z)|}{1 + |f(z)|^2} \leq c_1 \quad \text{for all } z \in B(0, r)$$

where c_1 denotes the constant $c_1 = \left[c(1-r^2)\right]^{-1}$.

Since f never takes the value 0, the function $t \to |f(tz)|$ is continuously differentiable for any fixed $z \in B(0,r)$ and

$$\left|\frac{d}{dt}(\arctan|f(tz)|)\right| \leq \frac{|f'(tz)||z|}{1+|f(tz)|^2} \leq c_1$$

so

$$\left|\arctan|f(z)| - \arctan|f(0)|\right| \leq \int_0^1 \left|\frac{d}{dt}(\arctan|f(tz)|)\right|dt \leq c_1$$

from which we get

$$\arctan|f(z)| \leq c_1 + \arctan|f(0)| \leq c_1 + \arctan M$$

which is the theorem with $C = \tan(c_1 + \arctan M)$. $\qquad\square$

From here we proceed exactly as before to prove Picard's big theorem etc.

Section 5 Exercises

1. Show that the metric ρ on $B(a,r)$, given by

$$\rho(z) := \frac{2r}{r^2 - |z-a|^2} \quad \text{for } z \in B(a,r)$$

has curvature -1.

2. Derive Schwarz' lemma from Ahlfors' lemma.

Hint: Consider the metric $f^*(\lambda_1)$ and note that

$$\frac{d}{dx}(\arcsin\phi(x)) = \frac{\phi'(x)}{1-\phi(x)^2}$$

3. In this exercise we shall give a proof of Liouville's theorem based on Ahlfors' lemma.

Let $f \in Hol(\mathbf{C})$ be bounded, say $f(\mathbf{C}) \subseteq B(0,r)$ for some $r \in]0,\infty[$.

(α) Show that $f^*(\lambda_r) \leq \lambda_R$ for any $R > 0$.

(β) Show that $f^*(\lambda_r) = 0$.

(γ) Derive Liouville's theorem.

4. In this exercise we give a proof of Picard's little theorem, based on Ahlfors' lemma and the metric ρ from Proposition 16(i). So let $f : \mathbf{C} \to \mathbf{C}\backslash\{0,1\}$ be holomorphic.

(α) Show that $f^*(\rho) \leq \lambda_R$ on any ball $B(0,R)$.

(β) Show that $f^*(\rho) = 0$.

(γ) Derive Picard's little theorem.

Chapter 8 Geometric Aspects and the Riemann Mapping Theorem

We start this chapter by studying geometric aspects of analytic maps, in particular of the so-called Möbius transformations. We finish by proving the Riemann mapping theorem which states that simply connected regions contained in $\mathbf{C}$ but different from $\mathbf{C}$, are conformally equivalent.

Section 1 The Riemann sphere

We have earlier studied analytic functions near isolated singularities. To handle such singularities it is aesthetically pleasing and often even convenient to introduce the Riemann sphere = the extended complex plane. As an example consider the map $z \to z^{-1}$ which has an isolated singularity at $z = 0$ and whose image does not contain 0. Adding the point ∞ to $\mathbf{C}$ and imposing the rules $\frac{1}{0} = \infty$ and $\frac{1}{\infty} = 0$ will make the map into a homeomorphism of the extended complex plane.

Definition 1.

The extended complex plane $\mathbf{C}_\infty$ consists of the complex plane plus an extra point ∞ (called infinity). We give $\mathbf{C}_\infty$ the topology which is defined by keeping the topology on $\mathbf{C}$ and by letting the sets $(\mathbf{C} \backslash K) \cup \{\infty\}$, where K runs through the compact subsets of $\mathbf{C}$, be the neighborhoods of ∞. The sets

$$B(\infty; r) := \{z \in \mathbf{C} \,|\, |z| > r\} \cup \{\infty\} \ , \ 0 < r < \infty$$

then constitute a neighborhood basis for the point ∞ .

It may be remarked that $\mathbf{C}_\infty$ is the usual 1-point compactification of $\mathbf{C}$ as found in any book on point set topology.

We adopt the following natural rules for computing with the symbol ∞ and a complex number z:

$$\frac{z}{\infty} = 0 \ , \ \ \infty + z = z + \infty = \infty \ \text{ for any } z \in \mathbf{C}$$
$$\frac{z}{0} = \infty \ , \ z \cdot \infty = \infty \cdot z = \infty \ \text{ for any } z \in \mathbf{C}\backslash\{0\}$$
$$0 \cdot \infty = \infty \cdot 0 = 0$$

We interpret $\mathbf{C}_\infty$ geometrically by exhibiting an explicit homeomorphism between the unit sphere

$$S^2 = \left\{ (x, y, z) \in \mathbf{R}^3 \,|\, x^2 + y^2 + z^2 = 1 \right\}$$

in $\mathbf{R}^3$ and the extended complex plane $\mathbf{C}_\infty$: As usual we identify the complex plane with the xy-plane in $\mathbf{R}^3$ such that the real axis is identified with the x-axis and the imaginary axis with the y-axis in $\mathbf{R}^3$. Stereographic projection from the north pole $N = (0,0,1)$ establishes a bijection π of $S^2 \backslash \{N\}$ onto $\mathbf{C}$. Putting $\pi(N) := \infty$ we see that π is a bijection of S^2 onto $\mathbf{C}_\infty$ which in coordinates is given by

$$\pi(x,y,z) = \begin{cases} \frac{x+iy}{1-z} & \text{for } (x,y,z) \in S^2 \backslash \{N\} \\ \infty & \text{for } (x,y,z) = N \end{cases}$$

Its inverse $\pi^{-1} : \mathbf{C}_\infty \to S^2$ has in coordinates the expression

$$\pi^{-1}(z) = \begin{cases} \dfrac{\left(2\,Re\,z, 2\,Im\,z, |z|^2 - 1 \right)}{|z|^2 + 1} & \text{for } z \in \mathbf{C} \\ (0,0,1) & \text{for } z = \infty \end{cases}$$

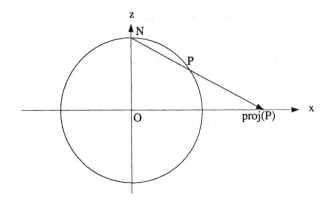

It is geometrically obvious (and also easily verified from the explicit expressions above) that both π and π^{-1} are continuous so that π indeed is a homeomorphism of S^2 onto $\mathbf{C}_\infty$.

We call S^2 for the *Riemann sphere* when we have in mind the connection between S^2 and $\mathbf{C}_\infty$ which is established above by stereographic projection from the north pole.

Definition 2.

A circle in $\mathbf{C}_\infty$ *is either an ordinary circle in* $\mathbf{C}$ *or a straight line in* $\mathbf{C}$ *with the point* ∞ *added.*

A pretty, geometric relation between S^2 and $\mathbf{C}_\infty$ is expressed by

Proposition 3.

Under the stereographic projection π *the circles of the sphere* S^2 *are mapped onto the circles of* $\mathbf{C}_\infty$.

Proof: The verification consists really just of elementary computations that are better left to the reader. We shall here only note that a circle on S^2 is the intersection between S^2 and a plane in $\mathbf{R}^3$. ☐

Let us return to the map $z \to z^{-1}$ which is a bijection of $\mathbf{C}_\infty$ onto itself. The corresponding bijection of the Riemann sphere onto itself is rotation by the angle π around the x-axis [seen by elementary computations from the formulas for π and π^{-1}]; in particular it maps neighborhoods of 0 onto neighborhoods of ∞ and vice versa. In the light of these remarks we introduce the following extensions of previous definitions :

Definition 4.

(i) A function $f : B(\infty; r) \to \mathbf{C}$ is said to be holomorphic around ∞ *if the function $w \to f(w^{-1})$ is holomorphic around 0.*

(ii) A function $f : B(\infty; r) \backslash \{\infty\} \to \mathbf{C}$ is said to have an isolated singularity at ∞ *if f is holomorphic in $B(\infty; r) \backslash \{\infty\}$.*

Such a singularity at ∞ is said to be removable / a pole of order k / an essential singularity *if $w \to f(w^{-1})$ has a removable singularity / a pole of order k / an essential singularity at 0.*

(iii) Let Ω be an open subset of $\mathbf{C}_\infty$. A function f which is holomorphic on Ω except for isolated singularities, all of which are poles, is said to be meromorphic in Ω. *More precisely, to each $z \in \Omega$ there must exist an open disc $B(z, r) \subseteq \Omega$ such that either f is holomorphic on $B(z, r)$ or f is holomorphic on the punctured disc $B(z, r) \backslash \{z\}$ and z is a pole.*

A holomorphic function is in particular meromorphic. In short, a meromorphic function is holomorphic except for poles. Essential singularities are not allowed. Meromorphic functions occur naturally as quotients between holomorphic functions: If f and g are holomorphic in an open subset Ω of $\mathbf{C}_\infty$ and if furthermore g does not vanish identically on any of the connected components of Ω, then f/g is meromorphic in Ω. We shall later see that the converse is true as well, i.e. any meromorphic function can be written as such a quotient (Exercise X.9). A rational function, i.e. a function of the form P/Q where P and Q are polynomials in z, is meromorphic in all of $\mathbf{C}_\infty$.

If f is meromorphic in an open subset Ω of $\mathbf{C}_\infty$ then we shall from now on ascribe the value ∞ to f at any point of Ω at which f has a pole. Then f is defined on all of Ω and is a continuous mapping of Ω into $\mathbf{C}_\infty$. In particular, any rational function is a continuous mapping of $\mathbf{C}_\infty$ into $\mathbf{C}_\infty$.

Section 2 The Möbius transformations

In this paragraph we introduce and study the Möbius transformations of the extended complex plane. They are important examples of rational maps with surprising and beautiful properties.

Definition 5.

A mapping $f : \mathbf{C}_\infty \to \mathbf{C}_\infty$ *of the form*

$$(1) \quad f(z) = \begin{cases} \frac{az+b}{cz+d} & \text{for } z \in \mathbf{C} \\ \frac{a}{c} & \text{for } z = \infty \end{cases}$$

where the four complex numbers $a, b, c, d \in \mathbf{C}$ *satisfy*

$$det \begin{Bmatrix} a & b \\ c & d \end{Bmatrix} = ad - bc \neq 0$$

is called a Möbius transformation *(other authors use terms like fractional linear transformation, bilinear transformation, homographic transformation or homography).*

Theorem 6.

The Möbius transformation (1) is a homeomorphism and a biholomorphic map of $\mathbf{C}_\infty$ *onto* $\mathbf{C}_\infty$.

If $c = 0$ *then* f *is a biholomorphic map of* $\mathbf{C}$ *onto* $\mathbf{C}$. *If* $c \neq 0$ *then* f *is a biholomorphic map of* $\mathbf{C}\backslash\left\{-\frac{d}{c}\right\}$ *onto* $\mathbf{C}\backslash\left\{-\frac{a}{c}\right\}$.

Proof: As is easy to check, the Möbius transformation

$$g(z) := \frac{ez + f}{gz + h} \quad \text{where} \quad \begin{Bmatrix} e & f \\ g & h \end{Bmatrix} = \begin{Bmatrix} a & b \\ c & d \end{Bmatrix}^{-1}$$

is the inverse of f. Being rational both f and g are continuous from $\mathbf{C}_\infty$ into $\mathbf{C}_\infty$.$\square$

The next two theorems are included simply because they are pretty. We won't need them later on.

Theorem 7.

If $A = \begin{Bmatrix} a & b \\ c & d \end{Bmatrix}$ *we let* f_A *denote the Möbius transformation*

$$f_A(z) := \frac{az + b}{cz + d} \quad \text{for } z \in \mathbf{C}_\infty$$

The map $A \to f_A$ *is a group-homomorphism of* $GL(2,\mathbf{C})$ *into the group of homeomorphisms of* $\mathbf{C}_\infty$. *So the Möbius transformations form a subgroup of the group of homeomorphisms of* $\mathbf{C}_\infty$.

Proof: Left to the reader. $\square$

Theorem 8.

The set of Möbius transformations equals the set of bijective meromorphic mappings of $\mathbf{C}_\infty$ *onto* $\mathbf{C}_\infty$.

Proof: The only statement which is not yet verified, is that any bijective meromorphic mapping $f : \mathbf{C}_\infty \to \mathbf{C}_\infty$ is a Möbius transformation. We verify it :

Since ∞ is either a pole or a removable singularity, f can near ∞ be written as

$$f(z) = z^N h\left(\frac{1}{z}\right) \text{ for some } N \in \{0, 1, 2, \cdots\}$$

where h is holomorphic near 0. In particular there exist constants C and R such that

$$|f(z)| \leq C|z|^N \text{ for } |z| \geq R$$

Since f has only finitely many singularities in $B[0, R]$ and since these are poles, there exists a polynomial $q \neq 0$ such that qf is entire. Combining this with the estimate above we get by Liouville's theorem that qf is a polynomial, say p. So $f = p/q$ is rational. We may of course assume that p and q have no 1ˢᵗ order factor in common.

If the degree of the polynomial p is bigger than 1, then p takes the value 0 at two different points or has a double root; in any case f is not injective (Cf. Theorem IV.21). Thus p is a polynomial of degree 1 or 0. Similar arguments reveal that q is of degree 1 or 0. So $f = p/q$ has the form

$$f(z) = \frac{az + b}{cz + d} \text{ where } a, b, c, d \in \mathbf{C}$$

If $det \begin{Bmatrix} a & b \\ c & d \end{Bmatrix} = 0$ then (a, b) is proportional to (c, d), so f reduces to a constant. But f was a bijection, so the determinant is not 0 . $\square$

Theorem 9.
Any Möbius transformation maps circles in $\mathbf{C}_\infty$ onto circles in $\mathbf{C}_\infty$.

Proof: Let us note that any Möbius transformation (1) is composed of rotations

$$z \to e^{i\theta} z \text{ where } \theta \in \mathbf{R}$$

dilations

$$z \to az \text{ where } a > 0$$

translations

$$z \to z + z_0 \text{ where } z_0 \in \mathbf{C}$$

and the inversion $z \to z^{-1}$. This is trivial if $c = 0$, and for $c \neq 0$ it is apparent from the easily verified formula

$$f(z) = \frac{a}{c} - \frac{ad - bc}{c} \frac{1}{cz + d}$$

Obviously rotations, dilations and translations map circles in $\mathbf{C}_\infty$ onto circles in $\mathbf{C}_\infty$, so it is just left to show that the inversion does too. But that is a consequence of Proposition 3 and the remark following it.

117

Here is an analytic proof for the reader who dislikes geometry: Lines and circles in $\mathbf{C}$ are sets of the form

$$\left\{ z \in \mathbf{C} \mid \alpha|z|^2 + \beta z + \overline{\beta}\overline{z} + \gamma = 0 \right\}$$

where α and γ range over $\mathbf{R}$ and β over $\mathbf{C}$ subject to the condition $|\beta|^2 > \alpha\gamma$. Replacement of z by $1/z$ transforms the equation

$$\alpha|z|^2 + \beta z + \overline{\beta}\overline{z} + \gamma = 0$$

into

$$\alpha + \beta\overline{z} + \overline{\beta}z + \gamma|z|^2 = 0$$

which is an equation of the same type. $\qquad\qquad\square$

Another important and useful property of Möbius transformations is that they preserve angles. To formulate that properly and to clarify the concepts we introduce a definition:

Definition 10.
Let $\alpha : I \to \mathbf{C}$ and $\beta : J \to \mathbf{C}$ be two paths and let $(t_0, s_0) \in I \times J$. Assume that $\alpha(t_0) = \beta(s_0))$ and that $\alpha'(t_0) \neq 0$ and $\beta'(s_0) \neq 0$.

The angle from α to β at (t_0, s_0) is the discrete set

$$\arg \beta'(s_0) - \arg \alpha'(t_0) = \arg \frac{\beta'(s_0)}{\alpha'(t_0)}$$

of real numbers.

Theorem 11.
Let Ω be an open subset of $\mathbf{C}$ and let $f : \Omega \to \mathbf{C}$ be complex differentiable at $z_0 \in \Omega$ with $f'(z_0) \neq 0$. Then f is angle preserving at z_0 in the following sense:

Let $\alpha : I \to \mathbf{C}$ and $\beta : J \to \mathbf{C}$ be two paths with $z_0 = \alpha(t_0) = \beta(s_0)$ and $\alpha'(t_0) \neq 0$ and $\beta'(s_0) \neq 0$.

Then the angle from $f \circ \alpha$ to $f \circ \beta$ at (t_0, s_0) equals the angle from α to β at (t_0, s_0).

Proof: The computation

$$\frac{(f \circ \beta)'(s_0)}{(f \circ \alpha)'(t_0)} = \frac{f'(\beta(s_0))\,\beta'(s_0)}{f'(\alpha(t_0))\,\alpha'(t_0)} = \frac{f'(z_0)\beta'(s_0)}{f'(z_0)\alpha'(t_0)} = \frac{\beta'(s_0)}{\alpha'(t_0)}$$

reveals that the proof is simpler than the statement. $\qquad\qquad\square$

Corollary 12.
The Möbius transformation (1) is angle preserving everywhere in $\mathbf{C} \setminus \left\{ -\frac{d}{c} \right\}$.

Proof: Differentiation of (1) yields

$$f'(z) = \frac{ad - bc}{(cz + d)^2} \neq 0$$

A more sophisticated argument runs as follows : f is injective (Theorem 6), hence $f'(z) \neq 0$ (Theorem IV.21). $\qquad\qquad$ □

The above results enable us to examine any given Möbius transformation by a minimal amount of computations, using in particular that Möbius transformations preserve circles in $\mathbf{C}_\infty$ and angles.

Example 13.

The Möbius transformation

$$f(z) := \frac{z - i}{z + i} \quad \text{for } z \in \mathbf{C}_\infty$$

occurs as the so-called *Cayley transform* in the theory of operators on a Hilbert space. Since f maps circles in $\mathbf{C}_\infty$ onto circles in $\mathbf{C}_\infty$ we get from $f(0) = -1$, $f(1) = -i$ and $f(\infty) = 1$ that $f(\mathbf{R} \cup \{\infty\})$ is the unit circle.

Combining the injectivity of f with $f(i) = 0$ and $f(-i) = \infty$ we conclude that f maps the upper half plane $\{z \in \mathbf{C} \mid \Im z > 0\}$ onto $B(0,1)$ and the lower half plane onto $B(\infty, 1)$. The mapping of the upper half plane onto the disc is also interesting from another point of view : It establishes a connection between two models of non-Euclidean geometry. In fact f is an isometry with respect to the non-Euclidean distances.

We proceed with another interesting example of a Möbius transformation that we shall need below in our proof of the Riemann mapping theorem.

Example 14.

Let $A(a; \cdot)$ for $a \in B(0,1)$ denote the Möbius transformation

$$A(a; z) := \frac{z - a}{1 - \bar{a} z} \quad \text{for } z \in \mathbf{C}_\infty$$

We claim that $A(a; \cdot)$ is a univalent map of $B(0,1)$ onto itself with $A(a; a) = 0$. And that any other such map is a constant (of modulus 1) multiple of $A(a; \cdot)$.

Note for later reference that $A(0; \cdot)$ is the identity map, that $A(-a; \cdot) = A(a; \cdot)^{-1}$, that $A'(a; a) = \left(1 - |a|^2\right)^{-1}$ and $A'(a; 0) = 1 - |a|^2$.

Proof of the claim : When $z \in S^1 = \partial B(0,1)$ then

$$|A(a; z)| = \left| \frac{z - a}{1 - \bar{a} z} \right| = |z| \left| \frac{1 - a z^{-1}}{1 - \bar{a} z} \right| = \left| \frac{1 - a \bar{z}}{1 - \bar{a} z} \right| = 1$$

so by the weak maximum principle $A(a; B(0, 1)) \subseteq B[0, 1]$, and further by the open mapping theorem $A(a; B(0, 1)) \subseteq B(0, 1)$.

Since the inverse mapping $A(a; \cdot)^{-1} = A(-a; \cdot)$ of $A(a; \cdot)$ is of the same form as $A(a; \cdot)$ we see that $A(a; \cdot)$ maps $B(0, 1)$ univalently onto $B(0, 1)$.

Let g be a univalent map of $B(0, 1)$ onto itself with $g(a) = 0$. Then both $f := g \circ \left\{ A(a; \cdot)^{-1} \right\}$ and its inverse satisfy the conditions of Schwarz' lemma, so $|f(z)| \leq |z|$ and $\left| f^{-1}(z) \right| \leq |z|$. But then

$$|f(z)| \leq |z| = \left| f^{-1}(f(z)) \right| \leq |f(z)|$$

so by the equality part of Schwarz' lemma there exists a constant $c \in \mathbf{C}$ with $|c| = 1$, such that $f(z) = cz$ for all $z \in B(0, 1)$. And then $g(z) = cA(a; z)$ for all $z \in B(0, 1)$.$\square$

Section 3 Montel's theorem

An important ingredient in our proof of the Riemann mapping theorem is the following compactness result which in the literature is often called the *condensation principle*. In functional analytic terms it states that the holomorphic functions on an open set constitute a Montel space.

Theorem 15 (Montel's theorem (1907)).

Let $\mathcal{F}$ be a sequence of holomorphic functions on an open subset Ω of $\mathbf{C}$, and assume that $\mathcal{F}$ is locally uniformly bounded in Ω.

Then $\mathcal{F}$ has a subsequence which converges locally uniformly in Ω to a holomorphic function.

The corresponding theorem is not true on $\mathbf{R}$ as the example $\{\sin nx \mid n = 1, 2, \cdots\}$ shows.

Proof: Let A be a countable dense subset of Ω. By the standard diagonal sequence argument we extract a subsequence $\{f_n\}$ from $\mathcal{F}$ which converges at each point $a \in A$. The program is to show that $\{f_n\}$ converges uniformly on each compact subset K of Ω. This will prove Montel's theorem because the limit function automatically will be analytic [by Weierstrass' theorem IV.10]. Since $C(K)$ is complete it suffices to show that $\{f_n|_K\}$ is a Cauchy sequence in $C(K)$. So let $\epsilon > 0$ and a compact subset K of Ω be given.

Since K is compact, $dist(K, \mathbf{C}\backslash\Omega) > 0$, so we can fix $R \in \,]0, dist(K, \mathbf{C}\backslash\Omega)[$, and let M be a bound for $\mathcal{F}$ on the compact subset $\{z \in \mathbf{C} \mid dist(z, K) \leq R\}$.

For $a \in \mathbf{C}$, $z \in K$, $|z - a| < R$ and $f \in \mathcal{F}$ we get by applying Schwarz' lemma to the function $w \to [f(z + wR) - f(z)]/2M$ that

$$|f(z) - f(a)| \leq 2M \frac{|z - a|}{R}$$

Now $a \in \mathbf{C}$, $z \in K$, $|z - a| < \min\left\{R, \frac{R\epsilon}{8M}\right\}$ and $f \in \mathcal{F}$ implies that $|f(z) - f(a)| < \epsilon/4$.

By the compactness of K we can cover it by finitely many discs $B(z_1, r), B(z_2, r), \cdots, B(z_s, r)$ with centers $z_1, z_2, \cdots, z_s$ in K and with (the same) radius $r = \min\left\{R, \frac{R\epsilon}{8M}\right\}/2$. Choose $a_j \in A \cap B(z_j, r)$ for $j = 1, 2, \cdots, s$. Then for any $z \in K$ there is an a_j such that

$$|z - a_j| < 2r = \min\left\{R, \frac{R\epsilon}{8M}\right\}$$

so by the estimate above

$$
\begin{aligned}
&|f_n(z) - f_m(z)| \\
&\leq |f_n(z) - f_n(a_j)| + |f_n(a_j) - f_m(a_j)| + |f_m(a_j) - f_m(z)| \\
&\leq \frac{\epsilon}{2} + |f_n(a_j) - f_m(a_j)| \\
&\leq \frac{\epsilon}{2} + \sum_{j=1}^{s} |f_n(a_j) - f_m(a_j)|
\end{aligned}
$$

which shows that

$$\|f_n - f_m\|_{\infty, K} \leq \frac{\epsilon}{2} + \sum_{j=1}^{s} |f_n(a_j) - f_m(a_j)|$$

The sequence $\{f_n(a_j)\}$ converges for each fixed j by the very choice of $\{f_n\}$, so $\|f_n - f_m\|_{\infty, K} \leq \epsilon$ when n and m both are sufficiently large. We have thus proved that $\{f_n|_K\}$ is a Cauchy sequence in $C(K)$ as desired.   □

The following useful result is a corollary of Montel's theorem. It says that analyticity is preserved when the parameter in a family of holomorphic functions is integrated away.

Proposition 16.

Let $f \in C(I \times \Omega)$ where I is an interval on the real line and Ω is an open subset of the complex plane. If

$$f(t, \cdot) \in Hol(\Omega) \text{ for each fixed } t \in I \ , \text{ and}$$

$$\sup_{z \in K} \int_I |f(t, z)| \, dt < \infty \text{ for each compact subset } K \text{ of } \Omega$$

then

$$z \to \int_I f(t, z) dt$$

is holomorphic on Ω.

Proof: If I is compact then Fubini and Morera secure the conclusion, so left is the non-compact case. Here we take an increasing sequence $\{I_n\}$ of compact subintervals I_n of I, such that their union is I. Then as just stated

$$F_n(z) := \int_{I_n} f(t,z)dt$$

is holomorphic on Ω. By the assumptions $\{F_n\}$ is a locally uniformly bounded sequence. It converges pointwise to

$$F(z) := \int_I f(t,z)dt$$

by the dominated convergence theorem. Montel's theorem implies that F is holomorphic on Ω. $\qquad\qquad\square$

Section 4 The Riemann mapping theorem

Definition 17.

Two open subsets V and W of the complex plane are said to be conformally equivalent *if there exists a* conformal equivalence $\phi : V \to W$, *i.e. a univalent map ϕ of V onto W.*

Conformal equivalence is an equivalence relation [by Theorem IV.20].

Examples 18:

1. The Cayley transformation

$$z \to \frac{z-i}{z+i}$$

is a conformal equivalence of the open upper half plane onto the unit disc.

2. $z \to e^z$ is a conformal equivalence of the strip $0 < \Im z < \pi$ onto the upper half plane and takes $i\pi/2$ into i. Combining with the Cayley transformation we see that

$$z \to \frac{1+ie^z}{e^z+i}$$

is a conformal equivalence of the strip onto the unit disc.

3. $z \to e^{-z}$ maps the strip $0 < \Im z < 2\pi$ conformally onto the plane cut along the negative real axis.

Definition 19.

A region *is an open, connected and non-empty subset of* C.

V and W are of course homeomorphic if they are conformally equivalent. So topological properties may prohibit domains to be conformally equivalent. For example, an annulus and a disc are not homeomorphic, and so a fortiori not conformally equivalent. But topology alone is not the complete story either: $\mathbf{C}$ and the unit disc $B(0,1)$ are homeomorphic, but $\mathbf{C}$ is not conformally equivalent to any bounded domain [Liouville's theorem], in particular not to the unit disc. So much the more chocking is the fantastic statement that B. Riemann enunciated in his Göttingen dissertation of 1851 :

Theorem 20 (The Riemann mapping theorem).
Every simply connected region other than $\mathbf{C}$ is conformally equivalent to the open unit disc.

The Riemann mapping theorem does not provide us with a formula for the conformal equivalence ϕ between the given simply connected domain V and the unit disc. For an explicit solution of a problem it will be necessary to have more information about ϕ than its mere existence. Explicit formulas for ϕ when V is bounded by polygons can be found in Section 17.6 of [HII]. For a discussion of an possible extension of ϕ to a homeomorphism of $\overline{V}$ onto B[0,1] see [No].

An easy, but surprising consequence of the Riemann mapping theorem is that any two simply connected regions of $\mathbf{C}$ are homeomorphic.

If $\phi : V \to B(0,1)$ is a conformal equivalence then $f \to f \circ \phi$ is an algebra isomorphism between $\mathrm{Hol}(B(0,1))$ and $\mathrm{Hol}(V)$. So any problem about the algebra $\mathrm{Hol}(V)$ can be carried over to a problem on the unit disc, and the (possible) solution then carried back from there to $\mathrm{Hol}(V)$.

We will prove the following version of the Riemann mapping theorem. The earlier one is of course a corollary.

Theorem 21 (The Riemann mapping theorem).
Any connected non-empty open set $\Omega \subseteq \mathbf{C}$ which is not the entire complex plane, and on which every never vanishing holomorphic function has a holomorphic square root, is conformally equivalent to the open unit disc. In fact, given $a \in \Omega$ there exists exactly one conformal equivalence $H : \Omega \to B(0,1)$ such that $H(a) = 0$ and $H'(a) > 0$.

Proof : For the sake of clarity we divide the proof into 5 steps. We let $\mathcal{J}$ be the set of all univalent maps of Ω into $B(0,1)$.

Step 1. We show that $\mathcal{J}$ is not empty.

Choose any $z_0 \in \mathbf{C} \backslash \Omega$ (recall that Ω is not all of $\mathbf{C}$). The function $z \to z - z_0$ does not vanish anywhere in Ω, so it has a holomorphic square root g:

$$[g(z)]^2 = z - z_0 \ \text{ for all } \ z \in \Omega$$

Clearly g must be univalent. It is also immediately verified that $-g(\Omega)$ and $g(\Omega)$ have empty intersection. $-g(\Omega)$ is open by the Open Mapping Theorem. Let $z_1 \in -g(\Omega)$

and $B(z_1, r) \subseteq -g(\Omega)$. The map $\psi(z) := r/(z - z_1)$ is univalent and maps the complement of $B(z_1, r)$ (in particular $g(\Omega)$) into the unit disc $B(0, 1)$. The composite map $\psi \circ g$ then belongs to $\mathcal{J}$. Thus $\mathcal{J}$ is not empty.

Step 2. We find H.

Every $g \in \mathcal{J}$ is univalent on Ω and therefore g' does not vanish on Ω. Thus

$$\eta := \sup_{g \in \mathcal{J}} |g'(a)| > 0$$

Choose $f_n \in \mathcal{J}$ for $n = 0, 1, 2, \cdots$ such that $\lim |f_n'(a)| = \eta$. The f_n are uniformly bounded on Ω so Montel's theorem applies. By choosing a subsequence if necessary we may assume that $\{f_n\}$ converges locally uniformly on Ω to a holomorphic function H. By Weierstrass' theorem (Theorem IV.10) $\{f_n'\}$ converges locally uniformly to H', so that $|H'(a)| = \eta$. In particular H is not constant.

Step 3. $H \in \mathcal{J}$.

Suppose not, so that there exist $x, y \in \Omega$, $x \neq y$ such that $H(x) = H(y)$. Choose a disc $B(x, r)$ whose closure is in Ω, that does not contain y and satisfies that

$$\inf \left\{ |H(z) - H(x)| \mid |z - x| = r \right\} > 0$$

This is possible because zeros of nonconstant holomorphic functions are isolated. If the sequence $\{f_n\}$ is as in step 2 we get for large enough n that

$$|f_n(x) - f_n(y)| < \inf \left\{ |f_n(z) - f_n(y)| \mid |z - x| = r \right\}$$

If $h_n(w) := f_n(w) - f_n(y)$ does not vanish in $B(x, r)$ then $1/h_n$ is a counterexample to the weak maximum principle. So

$$h_n(w) = f_n(w) - f_n(y) = 0 \text{ for some } w \in B(x, r)$$

But that contradicts that f_n is univalent.

Since $f_n(\Omega) \subseteq B(0, 1)$ and $\{f_n\}$ converges to H, we see that $H(\Omega) \subseteq B[0, 1]$. Since H is not a constant function $H(\Omega)$ is open. Thus $H(\Omega) \subseteq B(0, 1)$ and $H \in \mathcal{J}$.

Step 4. $H(a) = 0$ and H has range $B(0, 1)$.

Put now $\alpha = H(a)$ and let g be the composite map $g := A(\alpha, H)$. Then $g \in \mathcal{J}$. By a small computation

$$|g'(a)| = \frac{|H'(a)|}{1 - |\alpha|^2} = \frac{\eta}{1 - |\alpha|^2} > \eta \text{ unless } \alpha = 0$$

Thus $H(a) = 0$.

Suppose H is not onto $B(0, 1)$ and let $b \in B(0, 1) \backslash H(\Omega)$. The function $A(b, H)$ does not vanish in Ω, so it has a holomorphic square root ϕ in Ω: $[\phi(z)]^2 = A(b, H(z))$. Clearly $\phi \in \mathcal{J}$. Recalling that $H(a) = 0$ and $|H'(a)| = \eta$, we compute that

$$|\phi'(a)| = \eta \frac{1 - |b|^2}{2\sqrt{|b|}}$$

Now, $\psi := A(\phi(a), \phi)$ also belongs to $\mathcal{J}$, and we find that

$$|\psi'(a)| = \eta \frac{1 + |b|}{2\sqrt{|b|}} > \eta$$

which is a contradiction. Hence H is onto.

Step 5. The uniqueness.

The uniqueness of H is an easy consequence of the uniqueness statement in Example 14 above. $\square$

Section 5 Primitives

We have seen in Theorem V.8 that every holomorphic function on a simply connected domain has a primitive there. There are open sets for which this fails. For example the plane punctured at the origin has not got this property : The function $1/z$ defined on $C \backslash \{0\}$ does not possess a primitive [Because if it did, the line integral of $1/z$ along the unit circle would be zero, but it is $2\pi i$]. It is a most remarkable fact that a connected open set has this property if and only if it is homeomorphic to the open unit disc.

Theorem 22.

For any non-empty connected open set Ω the following are equivalent :

a) Every holomorphic function on Ω has a primitive.

b) Ω is homeomorphic to the unit disc.

Proof :

a) => b)

If $\Omega = C$, the map

$$z \to \left(1 - e^{-|z|}\right) \frac{z}{|z|}$$

is a homeomorphism of C onto the open unit disc. If $\Omega \neq C$ then the hypotheses of Theorem 21 are satisfied : Let f be a non-vanishing holomorphic function on Ω. By assumption f'/f has a primitive, say ϕ. It follows that $f = const\, e^{\phi}$. Adjusting ϕ by adding a suitable constant we see that f has a holomorphic logarithm, and in particular a holomorphic square root. By Theorem 21 Ω is conformally equivalent to the unit disc.

b) => a)

If Ω is homeomorphic to the unit disc then it is simply connected, and so we may appeal to Theorem V.8. $\square$

Section 6 Exercises

1. When is a Möbius transformation T a projection, i.e. when is $T^2 = T$?

125

2. Show that the Möbius transformation which takes -1 to 0, 0 to 1 and 1 to ∞, takes the unit circle (except 1) to the imaginary axis.

3. Find a Möbius transformation that maps the semi-disc $\{z \in B(0,1) \mid \Im z > 0\}$ onto the positive quadrant.

4. Find a Möbius transformation ϕ that maps the open right half plane onto $B(0,1)$ in such a way that $\phi(1) = 0$ and $\phi'(1) > 0$.

5. Let f and g be holomorphic mappings of $B(0,1)$ onto a subset Ω of the complex plane. Assume that f is a bijection of $B(0,1)$ onto Ω and that $f(0) = g(0)$. Show that

$$g(B(0,r)) \subseteq f(B(0,1)) \quad \text{for any} \ \ r \in \,]0,1[$$

6. Let $f : B(0,1) \to B(0,1)$ be holomorphic and fix more than one point. Show that $f(z) = z$ for all $z \in B(0,1)$.

7. (a) Why can't you display an analytic function $f : B(0,1) \to B(0,1)$ such that $f(1/2) = 3/4$ and $f'(1/2) = 2/3$?

(b) Can you find an analytic function $f : B(0,1) \to B(0,1)$ such that $f(0) = 1/2$ and $f'(0) = 3/4$? How many are there ?

8. Let π_+ be the open right half plane, and let $f : \pi_+ \to \pi_+$ be analytic. Assume that $f(a) = a$ for some $a \in \pi_+$. Show that $|f'(a)| \leq 1$.

9. Let $f,g : B(0,1) \to \mathbf{C}$ be holomorphic functions, g univalent. Assume $f(0) = g(0)$ and $f(B(0,1)) \subseteq g(B(0,1))$. Show that $|f'(0)| \leq |g'(0)|$ and that equality sign implies $f(B(0,1)) = g(B(0,1))$.

10. Prove Vitali-Porter's Theorem :

Theorem (Vitali 1903, Porter 1904).

Let Ω be a connected open subset of $\mathbf{C}$ and let $\{f_n\}$ be a sequence of holomorphic functions on Ω. Assume that the sequence $\{f_n\}$ is locally uniformly bounded in Ω, and that $\{f_n(z)\}$ has a limit as $n \to \infty$ for a set of $z \in \Omega$ with an accumulation point in Ω.

Then $\lim\limits_{n \to \infty} f_n(z)$ exists locally uniformly in Ω.

Hint : Montel's theorem.

11. Let $\{f_n\}$ be a locally uniformly bounded sequence of holomorphic functions in a connected open subset Ω of $\mathbf{C}$, and assume that f_n is zero-free in Ω for each n.

Show that $\lim f_n = 0$ locally uniformly in Ω if there exists a $z_0 \in \Omega$ such that $\lim f_n(z_0) = 0$.

Hint : Apply the maximum principle to $1/f_n$ and refer to the Vitali-Porter theorem.

12. In this problem we generalize a part of Schwarz' lemma.

Let Ω be a bounded, connected, open subset of $\mathbf{C}$, let $a \in \Omega$ and let $f : \Omega \to \Omega$ be a holomorphic function with $f(a) = a$.

We put

$$f_1 = f , \ f_{n+1} = f \circ f_n \ \text{ for } \ n = 1, 2, \cdots$$

(α) Show that there exists a constant K, depending on Ω and a but not on f, such that $|f'(a)| \leq K$.

(β) Show that $|f'(a)| \leq 1$. Hint : Compute $f_n'(a)$.

(γ) Assuming that $f'(a) = 1$, prove that $f(z) = z$ for all $z \in \Omega$. Hint: Write

$$f(z) - z = c_m(z - a)^m + (z - a)^{m+1}h(z)$$

and compute the coefficient c_m.

(δ) Assuming that $|f'(a)| = 1$, prove that f is a bijection of Ω onto Ω. Hint: Find a sequence of integers $n_k \nearrow \infty$ as $k \to \infty$ and a function g such that

$$f'(a)^{n_k} \to 1 \text{ and } f_{n_k} \to g$$

13. Show that $z \to e^{-z}$ is a conformal equivalence of the strip $-\pi < \Im z < \pi$ onto the plane cut along the negative real axis.

14. Find a conformal equivalence of

$$\{z \in \mathbf{C} \mid |z| < 1 \text{ and } |z - (1 + i)| < 1\}$$

onto the unit disc.

15. Find a conformal equivalence between the semi-disc

$$\{z \in B(0,1) \mid \Im z > 0\}$$

and the unit disc.

16. Find an explicit formula for that conformal equivalence H between

$$\{z \in \mathbf{C} \mid -1 < \Re z < 1\}$$

and $B(0,1)$ which satisfies $H(0) = 0$ and $H'(0) > 0$.

17. Show that $\{z \in \mathbf{C} \mid 0 < |z| < 1\}$ is not conformally equivalent to any annulus $\{z \in \mathbf{C} \mid r < |z| < R\}$ where $0 < r < R < \infty$. For more information on general annuli we refer the reader to Theorem 3 in Chapter 7, §1 of [Na].

18. Let Ω be an open connected subset of **C**. Let $\{f_n\}$ be a sequence of univalent analytic functions on Ω, and assume that $\{f_n\}$ converges locally uniformly to a function f.

Show that f is either 1-1 or a constant (Hurwitz 1889), and show that both possibilities can occur.

19. Let f be holomorphic and bounded in $\{z \in \mathbf{C} \mid -1 < \Im z < 1\}$ and assume that $f(x) \to 0$ as $x \to \infty$.

Prove that

$$\lim_{x \to \infty} f(x + iy) = 0 \text{ for each } y \in]-1, 1[$$

and that the passage to the limit is uniform when y is confined to any interval of the form $[-\alpha, \alpha]$ where $0 < \alpha < 1$. Hint: Consider $f_n(z) := f(z + n)$ for $z = x + iy$ in the square $\{x + iy \mid |x| < 1, |y| < 1\}$.

20. Show the following generalization of Schwarz' lemma:

Let f be analytic in a neighborhood of $B[0,1]$B[0,1] and let $|f(z)| \leq M$ for $|z| = 1$ and let $f(0) = a$. Then

$$|f(z)| \leq \frac{M|z| + |a|}{|a||z| + M} \text{ for } |z| < 1$$

Chapter 9 Meromorphic Functions and Runge's Theorems

Section 1 The argument principle

In this section we derive an important and beautiful result called the argument principle, which expresses the number of zeros of an analytic function inside a closed curve in terms of a winding number. More generally, it is a formula for the difference between the number of zeros and poles of a meromorphic function.

To reassure ourselves that we will not encounter meaningless expressions like $\infty - \infty$ we make the following observation.

Observation 1.

Let f be meromorphic in an open set $\Omega \subseteq C$, and assume that there is a compact subset of Ω that contains all the zeros (resp. poles) of f in Ω.

Then the number of zeros (resp. poles) of f in Ω is finite.

Proof : Left to the reader. □

We remind the reader that index = argument increment divided by 2π (Remark (3) following Definition III.12).

Theorem 2 (The argument principle).

Let Ω be a simply connected open subset of C, γ a closed curve in Ω and f meromorphic in Ω. Let $z_1, z_2, \cdots, z_r$ and $p_1, p_2, \cdots, p_s$ be all zeros and poles of f with multiplicities $n_1, n_2, \cdots, n_r$ and $m_1, m_2, \cdots, m_s$ respectively. Suppose that none of the zeros and poles lie on γ.

Then

$$Ind_{f \circ \gamma}(0) = \sum_{j=1}^{r} n_j \, Ind_\gamma(z_j) - \sum_{k=1}^{s} m_k \, Ind_\gamma(p_k)$$

If γ is even a closed path then

$$\frac{1}{2\pi i} \int_\gamma \frac{f'(z)}{f(z)} dz = \sum_{j=1}^{r} n_j \, Ind_\gamma(z_j) - \sum_{k=1}^{s} m_k \, Ind_\gamma(p_k)$$

See also Exercise 1.

Proof : Write

$$f(z) = \prod_{j=1}^{r} (z - z_j)^{n_j} \prod_{k=1}^{s} (z - p_k)^{-m_k} g(z)$$

where g is holomorphic and vanishes nowhere in Ω. Choose any continuous logarithms for $\gamma - z_j$, $\gamma - p_k$ and g (recall that g does not vanish in Ω). Then

$$\sum_{j=1}^{r} n_j \log (\gamma - z_j) - \sum_{k=1}^{s} m_k \log (\gamma - p_k) + \log (g \circ \gamma)$$

is a continuous logarithm for $f \circ \gamma$ The proof follows from the definition of Ind (Definition III.12) and the formula for the index of a path (Theorem III.17(d)). $\square$

To understand what the above means, let γ be a circle. For w inside γ the winding number $Ind_\gamma(w) = 1$ and for w outside it is zero. The argument principle thus says that the change in $\arg f(z)$ as z traces γ equals $Z - P$ where Z is the number of zeros and P the number of poles of f inside γ, multiplicities counted.

Example 3.

The argument principle can give information about the location of the zeros of a polynomial. We present here a typical example of the kind of problems that can be solved:

The polynomial $f(z) := z^8 + az^3 + bz + c$, where $a \geq 0$, $b \geq 0$ and $c > 0$ has exactly 2 zeros in the first quadrant.

If $t \geq 0$ then $f(t) > 0$ and $\Re(f(it)) > 0$, so f has no zeros on the positive semi-axes.

Let $R > 0$ and consider the quarter circle

$$z = R e^{i\theta} , \ 0 \leq \theta \leq \frac{\pi}{2} .$$

On this quarter circle f is

$$f\left(R e^{i\theta}\right) = R^8 e^{8i\theta}\left(1 + aR^{-5}e^{-5i\theta} + bR^{-7}e^{-7i\theta} + cR^{-8}e^{-8i\theta}\right)$$

For R sufficiently large the parenthesis has positive real part, so a continuous argument function for f along the quarter circle is

$$\theta \rightarrow 8\theta + Arg\left\{R^{-8}e^{-8i\theta} f\left(R e^{i\theta}\right)\right\}$$

where Arg denotes the principal determination of the argument.

Consider now the curve indicated on the following figure

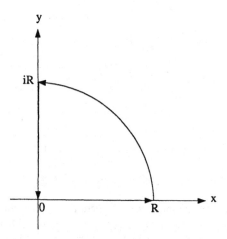

We have observed that $\Re(f(it)) > 0$ when $t \geq 0$. So as z varies from iR to 0 along the imaginary axis the argument of f increases by $Arg\, f(0) - Arg\, f(iR) = -Arg\, f(iR)$, and so the argument increment is in absolute value less than π.

The argument of f is constant (viz. equal to 0) from 0 to R. Finally from R to iR along the circle it increases by $\frac{8\pi}{2} + \Delta = 4\pi + \Delta$, where $|\Delta| < \pi$. So along the curve on the figure the argument of f increases by 4π + something , where $|$something$| < 2\pi$. But the argument increment is an integer multiple of 2π, so here it is exactly 4π.

By the argument principle there are exactly 2 zeros of f (counted with multiplicity) inside the contour, and hence (since $R > 0$ can be arbitrarily large) in the first quadrant.

Section 2 Rouché's theorem

The following clever lemma will be useful in the proofs of Rouché's and Runge's theorems. It is an integral representation theorem like the Cauchy integral formula. In contrast to the earlier results it asserts the existence of a curve with certain properties, not that every curve has those properties. The power of integral representations has already been demonstrated in connection with the Cauchy integral formula; it led to local power series expansions.

Lemma 4.

Let Ω be an open subset of $\mathbf{C}$ and let $F \subseteq \Omega$ be compact. Then there exists a finite set of piecewise linear closed paths $\gamma_1, \gamma_2, \cdots, \gamma_n$ in $\Omega \backslash F$ such that for any function

$f \in Hol(\Omega)$:

$$0 \quad = \quad \sum_{j=1}^{n} \int_{\gamma_j} f(z)dz \ \ and$$

$$f(a) = \frac{1}{2\pi i} \sum_{j=1}^{n} \int_{\gamma_j} \frac{f(z)}{z-a} dz \ \ for \ all \ \ a \in F$$

Proof :

Consider the following grid (i.e. set of squares) on $\mathbf{R}^2$:

$$\left\{ (x,y) \in \mathbf{R}^2 \mid \frac{i}{N} \leq x \leq \frac{i+1}{N} \ , \ \frac{j}{N} \leq y \leq \frac{j+1}{N} \right\}$$

where i and j range over the integers, and where the positive integer N is chosen so large that any square of the grid having a point in common with F is contained in Ω. We orient each of the squares in the counter clockwise direction.

Denote by $Q_1, Q_2, \cdots, Q_m$ the set of those squares in the grid that have a point in common with F. The union of these Q_i contains F - that much is clear. Note also that if for example a vertex of a square belongs to F, then all the four squares having this point as a vertex are included in the system $Q_1, Q_2, \cdots, Q_m$. A similar remark applies to a point on an edge. Let $L_1, L_2, \cdots, L_r$ denote the boundary segments of this system of squares, i.e. $L_1, L_2, \cdots, L_r$ are those edges of the system which are edges of exactly one square. Then L_j must for each j be contained in $\Omega \backslash F$. Since each square is oriented, each segment is oriented, too.

Now let f be holomorphic in Ω. Since $Q_k \subseteq \Omega$ we get by Cauchy's theorem that

$$(1) \quad \int_{\partial Q_k} f(z)dz = 0 \ \ for \ \ 1 \leq k \leq m$$

And if a belongs to the interior of the square Q_k (it can belong to only one)

$$(2) \quad \frac{1}{2\pi i} \int_{\partial Q_k} \frac{f(z)}{z-a} dz = f(a) \ , \ \ a \in Int(Q_k)$$

and for the other squares

$$(3) \quad \frac{1}{2\pi i} \int_{\partial Q_l} \frac{f(z)}{z-a} dz = f(a) \ , \ \ a \in Int(Q_k) \, , \ k \neq l$$

Adding (1) over all k we find that

$$\sum_k \int_{\partial Q_k} f(z)dz = 0$$

If an edge is common to two squares, then the orientations induced on it are opposite and so the net contribution is zero. We get

$$\sum_j \int_{L_j} f(z)dz = 0$$

Using the same reasoning we get from (2) and (3) that if a belongs to the interior of one of the squares then

$$(4) \quad \sum_j \frac{1}{2\pi i} \int_{L_j} \frac{f(z)}{z-a}dz = f(a)$$

Now every $b \in F$ is a positive distance from $L_1 \cup L_2 \cup \cdots \cup L_r$ and is a limit of a's that belong to the interior of one of the squares. (4) is therefore by continuity valid for all $a \in F$.

We shall finally show that the oriented line segments

$$L_j = [a_j, b_j] \quad \text{where} \quad j = 1, 2, \cdots, r$$

can be grouped together to a collection of closed paths $\gamma_1, \gamma_2, \cdots, \gamma_n$.

For any polynomial p we get from what we have already shown that

$$0 = \sum_{j=1}^r \int_{L_j} p'(z)dz = \sum_{j=1}^r \{p(b_j) - p(a_j)\}$$

so

$$(5) \quad \sum_{j=1}^r p(b_j) = \sum_{j=1}^r p(a_j) \quad \text{for any polynomial } p$$

For $z_0 \in \mathbb{C}$ let

$$n_a := \text{the number of j for which } a_j = z_0$$
$$n_b := \text{the number of j for which } b_j = z_0$$

Choose a polynomial p such that $p(z_0) = 1$ and such that $p = 0$ at the other points, i.e. at

$$\{a_1, a_2, \cdots, a_r, b_1, b_2, \cdots, b_r\} \backslash \{z_0\}$$

We conclude that the a's are the same as the b's; more precisely that there exists a permutation π of $\{1, 2, \cdots, r\}$ such that $b_j = a_{\pi(j)}$ for $j = 1, 2, \cdots, r$, and so $L_j = [a_j, a_{\pi(j)}]$ for $j = 1, 2, \cdots, r$.

Any permutation is a product of disjoint cycles, and each of the cycles of π gives rise to a closed path of L_j's. $\square$

Lemma 5.

Let Ω be open, let $F \subseteq \Omega$ be compact and let f be meromorphic in Ω. Suppose all the zeros and poles of f are in F. Then if the γ_j are as in Lemma 4 we have

$$\frac{1}{2\pi i} \sum_j \int_{\gamma_j} \frac{f'(z)}{f(z)} dz = Z - P$$

where Z and P are the number of zeros and poles of f, counted with multiplicities.

Proof : The proof is similar to that of Theorem 2. Represent f as there with the z_i and p_j as the zeros and poles of f with the corresponding multiplicities n_i and m_j. Differentiating we find

$$\frac{f'}{f} = \frac{g'}{g} + \sum_i \frac{n_i}{z - z_i} - \sum_j \frac{m_j}{z - p_j}$$

Lemma 4 applied to the holomorphic function $f = 1$ gives

$$\frac{1}{2\pi i} \sum_j \int_{\gamma_j} \frac{dz}{z - a} = 1 \text{ for all } a \in F$$

Since g'/g is holomorphic in Ω we get the desired result from Lemma 4. $\qquad \square$

The following result (Rouché's theorem) has many applications as is apparent from the exercises to this chapter. Another version of it was proved in Chapter IV (Theorem IV.13).

Theorem 6 (Rouché's theorem).

Let K be a compact subset of $\mathbf{C}$ with interior $Int(K)$. Let the two mappings $f, g : K \to \mathbf{C}_\infty$ be meromorphic in $Int(K)$, continuous on K, finite-valued on ∂K and there satisfying

$$(*) \quad |f(z) - g(z)| < |f(z)| + |g(z)| \text{ for all } z \in \partial K$$

Let $Z(f)$ and $P(f)$ (resp. $Z(g)$ and $P(g)$) denote the number of zeros and poles of f (resp. g) in $Int(K)$, counted with multiplicity.

Then $Z(f)$, $P(f)$, $Z(g)$ and $P(g)$ are finite, and $Z(f) - P(f) = Z(g) - P(g)$.

The crucial condition (*) expresses the natural condition that f and g should be "close" to one another, if only on ∂K.

Proof : By (*) neither f nor g has zeros on ∂K. Since they are finite-valued there, we see that there is a neighborhood $U \subseteq \mathbf{C}$ of ∂K such that f and g are finite-valued on $U \cap K$ and the inequality (*) is still true for all $z \in U \cap K$. In particular all the poles and zeros of f and g belong to the compact subset $F := K \backslash U$ of $\text{Int}(K)$. Hence their numbers are finite (Observation 1 above).

Let us for $\lambda \in [0,1]$ consider the meromorphic function $h_\lambda := \lambda f + (1 - \lambda)g$, which has all its poles and zeros off U, i.e. in F. By Lemma 5 we see that the quantity

$$\frac{1}{2\pi i} \sum_j \int_{\gamma_j} \frac{h'_\lambda(z)}{h_\lambda(z)} dz$$

is integer for each $\lambda \in [0,1]$. Being continuous as a function of λ it takes the same value at $\lambda = 0$ and at $\lambda = 1$. □

A special case of Rouché's theorem is the following :

Theorem 7 (Rouché's theorem).

Let f and g be holomorphic in an open set containing a compact set K. If

$$|f(z) - g(z)| < |f(z)| + |g(z)| \quad \text{for all } z \in \partial K$$

then f and g have the same (finite) number of zeros in K.

Section 3 Runge's theorems

We know that a function which is holomorphic in a disc, can be expanded in a power series, and so be approximated uniformly by polynomials on any compact subdisc. If a function is holomorphic in an annulus, say in $B(0,1) \backslash \{0\}$, then it cannot necessarily be approximated by polynomials. The obvious counter example is $z \to z^{-1}$. But we can in an annulus resort to a Laurent expansion and hence approximate it by rational functions. Runge's theorems deal with approximation of holomorphic functions by polynomials or rational functions on more general sets than discs and annuli.

We remind the reader that we have introduced the concept of a pole at ∞ (Definition VIII.4). To set the stage for Runge's theorem on approximation by rational functions we note that a polynomial (non-constant to be pedantic) is a rational function with a pole at ∞. And that a rational function whose only pole is at ∞, is a polynomial.

Theorem 8 (Runge's theorem on approximation by rational functions).

Let F be a compact subset of $\mathbf{C}$. Let $S \subseteq \mathbf{C}_\infty$ be a set which meets each connected component of $\mathbf{C}_\infty \backslash F$. Let f be holomorphic in an open set containing F.

Then to each $\epsilon > 0$ there exists a rational function P/Q with poles only in S such that

$$\left| f(z) - \frac{P(z)}{Q(z)} \right| < \epsilon \quad \text{for all } z \in F$$

Proof : In the proof it will be convenient to interpret expressions of the form $P\left(\frac{1}{z-\infty}\right)$, where P is a polynomial, as polynomials in z.

Letting Ω denote the open set in question there is according to Lemma 4 a finite set $\gamma_1, \gamma_2, \cdots, \gamma_n$ of oriented segments in $\Omega \backslash F$ such that

$$(5) \quad f(z) = \frac{1}{2\pi i} \sum_{j=1}^{n} \int_{\gamma_j} \frac{f(w)}{w - z} dw \quad \text{for all } z \in F$$

so the theorem will be proved once we show that each summand on the right hand side of (5) can be approximated uniformly on F by finite sums of the type $P\left(\frac{1}{z-c}\right)$, where the P's are polynomials and the c's belong to S. Let us just treat the first summand; the other terms may be handled in exactly the same way.

The line integral

$$\int_{\gamma_1} \frac{f(w)}{w - z} dw$$

can be uniformly approximated by Riemann sums

$$\sum_{j} \frac{A_j}{b_j - z} \quad \text{where } b_j \in \gamma^* .$$

Again let us just treat the first term $(b_1 - z)^{-1}$, the rest being handled similarly.

We drop the subscript and write b for b_1. γ_1 does not meet F so $b \notin F$. Let $c \in S$ be in the same connected component of $\mathbf{C}_\infty \backslash F$ as b, and assume first that $c \neq \infty$. Join b and c by a curve contained entirely in this component. Since this is disjoint from F, there is a positive distance between this curve and F. We can find points $b = c_0, c_1, \cdots, c_m = c$ on the curve such that

$$(6) \quad 2|c_i - c_{i+1}| < dist\,(curve, F)$$

Then writing $z - b = z - c_1 + c_1 - b$ we get

$$\frac{1}{z - b} = (z - c_1)^{-1} \sum_{k=0}^{\infty} (b - c_1)^k (z - c_1)^{-k}$$

the series converging uniformly for $z \in F$ by (6). So we may choose N so that

$$(7) \quad \left| (z - b)^{-1} - \sum_{k=0}^{N} (b - c_1)^k (z - c_1)^{-k-1} \right|$$

is uniformly small for $z \in F$. We can use the same argument on each term of the sum in (7) to replace c_1 by c_2. Namely

$$(z - c_1)^{-k} = (z - c_2 + c_2 - c_1)^{-k}$$

which can be expanded in a series that - because of (6) - converges uniformly for $z \in F$. In other words we can find a polynomial in $(z - c_2)^{-1}$ which approximates $(z - b)^{-1}$ uniformly on F. This procedure can be repeated until we reach $c = c_m$.

The argument is thus complete except when the point c chosen in S is ∞. To treat this case consider $(z - b)^{-1}$ where b belongs to the unbounded component of $\mathbf{C}\backslash F$. Choose d in the same component such that $\left|\frac{c}{d}\right| < \frac{1}{2}$ for all $z \in F$. From what we have done above there is a polynomial P such that $P\left(\frac{1}{z-d}\right)$ approximates $(z - b)^{-1}$ uniformly on F. Since $\left|\frac{c}{d}\right| < \frac{1}{2}$ each $(z - d)^{-1}$ occurring in the polynomial can be expanded in powers of z/d, i.e. there is a polynomial which approximates $(z - b)^{-1}$ uniformly on F.

The argument is thus complete. $\qquad\qquad\qquad\qquad\qquad\qquad\qquad\qquad\square$

For the next result (Runge's polynomial approximation theorem) and for the discussion of the inhomogeneous Cauchy-Riemann equation we need a fact from point set topology:

Proposition 9.

Let Ω be an open subset of the complex plane. Then there exists an increasing sequence $K_1, K_2, \cdots$ of compact subsets of Ω such that

(a) $K_1 \cup K_2 \cup \cdots = \Omega$

(b) $K_n \subseteq Int\, K_{n+1}$ for all $n = 1, 2, \cdots$

(c) Every compact subset of Ω is contained in K_n for some n

(d) Every connected component of $\mathbf{C}_\infty \backslash K_n$ contains a connected component of $\mathbf{C}_\infty \backslash \Omega$ for $n = 1, 2, \cdots$

The last statement (d) says that K_n has no holes except the ones coming from the holes of Ω (make a drawing!).

Proof :

The cases $\Omega = \emptyset$ and $\Omega = \mathbf{C}$ are trivial, so we shall concentrate on the remaining case in which $\partial\Omega \neq \emptyset$.

We will show that we for $n = 1, 2, \cdots$ may take

$$K_n := \left\{ z \in \mathbf{C} \,|\, dist(z, \partial\Omega) \geq \frac{1}{n} \right\} \cap B[0, n]$$

This K_n is a closed and bounded subset of $\mathbf{C}$, so it is compact. It is now almost obvious that (a), (b) and (c) are satisfied, so left is only point (d).

Here we begin by noting that if a connected component of $\mathbf{C}_\infty \backslash K_n$, say U, intersects a connected component of $\mathbf{C}_\infty \backslash \Omega$, say V, then U contains V: Indeed, since $\mathbf{C}_\infty \backslash K_n \supseteq \mathbf{C}_\infty \backslash \Omega$ the connected components of $\mathbf{C}_\infty \backslash K_n$ cover $\mathbf{C}_\infty \backslash \Omega$; in particular they cover V. They are open and disjoint, so by the connectedness of V we get $U \supseteq V$.

Hence (d) boils down to proving that each connected component U of $\mathbf{C}_\infty \backslash K_n$ intersects $\mathbf{C}_\infty \backslash \Omega$. To do so we notice that

$$\mathbf{C}_\infty \backslash K_n = \bigcup_{z \notin \Omega} B\left(z, \frac{1}{n}\right) \cup B(\infty, n)$$

Let $w \in U$. Then w lies in one of the balls, say $w \in B\left(z_0, \frac{1}{n}\right)$ where $z_0 \notin \Omega$. The ball is a connected subset of $\mathbf{C}_\infty \backslash K_n$, so $B\left(z_0, \frac{1}{n}\right) \subseteq U$, U being a connected component. Now, $z_0 \in U \cap (\mathbf{C}_\infty \backslash \Omega)$, so U intersects $\mathbf{C}_\infty \backslash \Omega$. $\qquad\square$

Theorem 10 (Runge's polynomial approximation theorem).

Let Ω be an open subset of $\mathbf{C}$. Every holomorphic function on Ω can be approximated uniformly on compacts by polynomials iff $\mathbf{C}_\infty \backslash \Omega$ is connected.

Proof: We will first show that the condition is sufficient. Proposition 9 tells (c) that it suffices to approximate the given holomorphic function f on any given K_n, (d) that $\mathbf{C}_\infty \backslash K_n$ has only one connected component. That one contains $\mathbf{C}_\infty \backslash \Omega$ and so it contains ∞. Take now $S = \{\infty\}$ and $F = K_n$ in Runge's theorem on approximation by rational functions (Theorem 8) to get polynomials to approximate f uniformly on K_n.

Now for the necessity: Suppose that $\mathbf{C}_\infty \backslash \Omega$ is not connected. Then there exist two closed, disjoint, non-empty sets F and F_1 with union $\mathbf{C}_\infty \backslash \Omega$. We may suppose that $\infty \in F_1$ so that F is a compact subset of $\mathbf{C}$. Clearly

$$\mathbf{C}_\infty \backslash F_1 = \Omega \cup F$$

is open. Put $O := \Omega \cup F$. F is a compact subset of the open set O. Let $\gamma_1, \gamma_2, \cdots, \gamma_n$ be paths in $O \backslash F = \Omega$ as in Lemma 4. Let $a \in F$. The function $z \to (z-a)^{-1}$ is holomorphic on Ω.

Assume now that all holomorphic functions on Ω are uniform limits on compacta of polynomials. Then we can in particular approximate the function $z \to (z-a)^{-1}$ by a sequence $\{P_k\}$ of polynomials on the union of the paths $\gamma_1, \gamma_2, \cdots, \gamma_n$. These polynomials are of course holomorphic in O and hence by the first part of the conclusion of Lemma 4 we have

$$\sum_j \int_{\gamma_j} P_k(z) dz = 0$$

which in the limit as $k \to \infty$ becomes

$$\sum_j \int_{\gamma_j} \frac{1}{z-a} dz = 0$$

But that contradicts the second part of Lemma 4. $\qquad\square$

Remark 11. *If Ω be an open subset of $\mathbf{C}$, then $\mathbf{C}_\infty \backslash \Omega$ is connected iff Ω is simply connected.*

Proof: Suppose first that $\mathbf{C}_\infty \backslash \Omega$ is connected. Any function f which is holomorphic on Ω may by Runge's polynomial approximation theorem be approximated uniformly by polynomials on compact subsets of Ω. So if γ is a closed path in Ω we conclude that $\int_\gamma f dz = 0$ which implies that f has a primitive (use the proof of Lemma II.12) and hence that Ω is simply connected (Theorem VIII.22).

Assume conversely that $\mathbf{C}_\infty \backslash \Omega$ is not connected. We then want to show that Ω is not simply connected. We assume it is and arrive at a contradiction as follows: Let F and F_1 be closed, disjoint, non-empty sets with union $\mathbf{C}_\infty \backslash \Omega$. Without loss of generality we may suppose that $\infty \in F_1$ so that F is a compact subset of $\mathbf{C}$. Clearly

$$\mathbf{C}_\infty \backslash F_1 = \Omega \cup F$$

is open. Put $O := \Omega \cup F$. F is a compact subset of the open set O. Let $\gamma_1, \gamma_2, \cdots, \gamma_n$ be the closed paths in $O \setminus F = \Omega$ from Lemma 4. Let $a \in F$. Since $z \rightarrow (z-a)^{-1}$ is holomorphic in Ω we get according to The Global Cauchy Theorem (Theorem V.7) that

$$\sum_j \int_{\gamma_j} \frac{1}{z-a} dz = 0$$

which contradicts the second part of Lemma 4. □

Remark 12. Let K be a compact set. Then the unbounded component of $\mathbf{C}\backslash K$ is precisely those $a \in \mathbf{C}$ for which the function $z \rightarrow (z-a)^{-1}$ is a uniform limit on K of polynomials (This follows from the proof of Theorem 8).

Runge's theorem on polynomial approximation is not the last word in these matters. Clearly, if a function f on a compact set K is the uniform limit of a sequence of polynomials then $f \in C(K) \cap Hol(Int(K))$, so this is a necessary condition for approximation by polynomials. Mergelyan's theorem states that it is also sufficient:

Theorem 13 (Mergelyan's theorem (1952)).
Let K be a compact subset of $\mathbf{C}$ with $\mathbf{C}\backslash K$ connected. Let $f \in C(K) \cap Hol(Int(K))$ Then there exists a sequence of polynomials converging to f uniformly on K.

Note that when K has empty interior then the only condition remaining on f is that $f \in C(K)$. So the classical approximation theorem of Weierstrass for an interval is a particular case of Mergelyan's theorem. For a comprehensible proof of Mergelyan's theorem the reader can consult Chapter 20 of [Ru].

Another extension of Weierstrass' approximation theorem is provided by

Theorem 14 (Carleman's theorem (1927)).
If $f \in C(\mathbf{R})$ then there exists to every continuous function $\epsilon : \mathbf{R} \rightarrow]0, \infty[$ an entire function F with the property that

$$|f(x) - F(x)| < \epsilon(x) \text{ for all } x \in \mathbf{R}.$$

139

A proof can be found in [Bu;Chap.VIII,§5]. For further generalizations (Arakelian's theorem, Keldysh-Lavrent'ev's theorem) see the references in the notes to Chapter VIII of [Bu], the note [Ga], [RR] and the survey article [Vi].

Section 4 The inhomogeneous Cauchy-Riemann equation

In this section we shall use Runge's theorem to show that the inhomogeneous Cauchy-Riemann equation has a solution on any open subset Ω of $\mathbf{C}$. We recall from Chapter II that the Cauchy-Riemann operator is the first order differential operator

$$\frac{\partial}{\partial \bar{z}} := \frac{1}{2}\left(\frac{\partial}{\partial x} + i\frac{\partial}{\partial y}\right) \ \text{ on } \ \mathbf{C} = \mathbf{R}^2$$

and that a function $u \in C^1(\Omega)$ is holomorphic on Ω iff it satisfies the homogeneous Cauchy-Riemann equation

$$\frac{\partial u}{\partial \bar{z}} = 0$$

on Ω. In the case of $\Omega = \mathbf{C}$ an explicit formula for a solution can be displayed:

Proposition 15.
If $f \in C_0^\infty(\mathbf{R}^2)$ and

$$u(z) := \frac{1}{\pi}\int\limits_{\mathbf{R}^2} \frac{f(z-w)}{w}d\mu(w) \ \text{ for } \ z \in \mathbf{C}$$

where $d\mu$ denotes the Lebesgue measure on $\mathbf{R}^2$, then $u \in C^\infty(\mathbf{R}^2)$ and $\partial u/\partial \bar{z} = f$.

Although the function $w \to w^{-1}$ has a pole at $w = 0$ it is nevertheless integrable over any ball, so that the integral in the proposition makes sense: Indeed, using polar coordinates we get

$$\int\limits_{B(0,1)} \frac{d\mu(w)}{|w|} = \int\limits_0^1\int\limits_0^{2\pi} \frac{1}{r}r d\phi dr = 2\pi < \infty$$

For any constant coefficient differential operator $D \neq 0$ on $\mathbf{R}^n$ the equation $D u = f \in C_0^\infty(\mathbf{R}^n)$ has a solution $u \in C^\infty(\mathbf{R}^n)$: That is known from the theory of partial differential equations. We may for example choose $u = f * E$, where E is a fundamental solution of D. A fundamental solution of the Cauchy-Riemann operator $\partial/\partial \bar{z}$ is $E(z) = 1/(\pi z)$; that produces the formula of the proposition. We will, however, not appeal to the theory of partial differential equations, but prove directly that the formula of the proposition works.

Proof of Proposition 15:

Since the function $w \to w^{-1}$ is locally integrable we see that u is continuous and that we may differentiate u by doing so under the integral sign. So for any differential operator D with constant coefficients we get

$$(Du)(z) = \frac{1}{\pi} \int_{\mathbf{R}^2} \frac{(Df)(z - w)}{w} d\mu(w)$$

which shows that $u \in C^\infty(\mathbf{R}^2)$.

To prove the other statement we fix $z \in \mathbf{C}$. For simplicity in writing we delete the subscript $\mathbf{R}^2$ on the integrals and put $F(w) := f(z_0 - w)$. Now,

$$\frac{\partial u}{\partial \overline{z}}(z_0) = \frac{1}{\pi} \int \frac{\partial f}{\partial \overline{z}}(z_0 - w) \frac{d\mu(w)}{w}$$
$$= -\frac{1}{\pi} \int \frac{\partial}{\partial \overline{w}} \{f(z_0 - w)\} \frac{d\mu(w)}{w} = -\frac{1}{\pi} \int \frac{\partial F}{\partial \overline{w}}(w) \frac{d\mu(w)}{w}$$

Using that $\partial / \partial \overline{w}$ is of first order (so that the usual formula for differentiating a product holds) and that $w \to w^{-1}$ is holomorphic, we get

$$\frac{\partial}{\partial \overline{w}} \left\{ \frac{F(w)}{w} \right\} = \frac{\partial F}{\partial \overline{w}}(w) \frac{1}{w} + F(w) \frac{\partial(w^{-1})}{\partial \overline{w}} = \frac{\partial F}{\partial \overline{w}}(w) \frac{1}{w}$$

so

$$\frac{\partial u}{\partial \overline{z}}(z_0) = -\frac{1}{\pi} \int \frac{\partial}{\partial \overline{w}} \left\{ \frac{F(w)}{w} \right\} d\mu(w) = -\frac{1}{\pi} \lim_{\epsilon \to 0} \int_{|w| > \epsilon} \frac{\partial G}{\partial \overline{w}}(w) d\mu(w)$$

where we, again for the sake of simplicity, use the notation $G(w) := F(w)/w$. Introducing the definition of the Cauchy-Riemann operator we find by a small computation that

$$2 \frac{\partial G}{\partial \overline{w}} = div(G, iG)$$

so Green's theorem for the annulus $\epsilon < |w| < R$, where R is chosen so large that G vanishes out there, implies that

$$\frac{\partial u}{\partial \overline{z}} = \frac{1}{2\pi} \lim_{\epsilon \to 0} \int_{|w| = \epsilon} (G(w), iG(w)) \cdot \left(\frac{\Re w}{|w|}, \frac{\Im w}{|w|} \right) d\lambda(w)$$

141

where $d\lambda$ denotes the usual arc length measure on the circle $|w| = \epsilon$. Now,

$$
\begin{aligned}
\frac{\partial u}{\partial \overline{z}} &= \frac{1}{2\pi} \lim_{\epsilon \to 0} \left\{ \frac{1}{\epsilon} \int\limits_{|w|=\epsilon} [G(w)\Re w + iG(w)\Im w] d\lambda(w) \right\} \\
&= \frac{1}{2\pi} \lim_{\epsilon \to 0} \left\{ \frac{1}{\epsilon} \int\limits_{|w|=\epsilon} G(w) w \, d\lambda(w) \right\} \\
&= \frac{1}{2\pi} \lim_{\epsilon \to 0} \left\{ \frac{1}{\epsilon} \int\limits_{|w|=\epsilon} F(w) d\lambda(w) \right\} \\
&= \lim_{\epsilon \to 0} \left\{ \frac{1}{2\epsilon\pi} \int\limits_{|w|=\epsilon} [F(w) - F(0)] d\lambda(w) + F(0) \right\} \\
&= 0 + F(0) = F(0) = f(z_0)
\end{aligned}
$$

$\square$

Theorem 16.

Let Ω be an open subset of the complex plane. The inhomogeneous Cauchy-Riemann equation $\partial u / \partial \overline{z} = f$ has a solution $u \in C^{\infty}(\Omega)$ for any $f \in C^{\infty}(\Omega)$.

Proof:

Let $K_0 = \emptyset \subseteq K_1 \subseteq K_2 \subseteq \cdots$ be an increasing sequence of compact subsets of Ω as in Proposition 9.

Note: Any function which is holomorphic on an open set containing K_n, can be approximated uniformly on K_n by elements from $\text{Hol}(\Omega)$. (Combine Runge's theorem (Theorem 8) with the property (d) of Proposition 9).

Choose $\phi_1, \phi_2, \cdots \in C_0^{\infty}(\Omega)$ such that $\phi_n = 1$ on K_n.

By Proposition 15 there exists $u_1 \in C^{\infty}(\Omega)$ (even in $C^{\infty}(\mathbf{R}^2)$) such that $\partial u_1 / \partial \overline{z} = \phi_1 f$, and $u_2 \in C^{\infty}(\Omega)$ such that $\partial u_2 / \partial \overline{z} = \phi_2 f$. Since $\phi_1 = \phi_2 (= 1)$ on K_1 we get

$$
\frac{\partial(u_2 - u_1)}{\partial \overline{z}} = 0 \quad \text{on} \ K_1
$$

so $u_2 - u_1$ is holomorphic on $\text{Int}(K_1)$. Trivially

$$
|u_2 - u_1| < \frac{1}{2} \quad \text{on} \ K_0
$$

because $K_0 = \emptyset$.

142

Section 4. The inhomogeneous Cauchy-Riemann equation

Assume now that we have produced $u_1, u_2, \cdots, u_n \in C^\infty(\Omega)$ such that

$$\frac{\partial u_j}{\partial \bar{z}} = \phi_j f \ ,$$

$$u_{j+1} - u_j \in Hol(Int\, K_j) \ \text{and}$$

$$|u_{j+1} - u_j| < \frac{1}{2^j} \ \text{on } K_{j-1} \ \text{for } j = 1, 2, \cdots, n-1$$

We will extend the system $\{u_1, u_2, \cdots, u_n\}$ by adding one function more: Choose $v \in C^\infty(\Omega)$ such that

$$\frac{\partial v}{\partial \bar{z}} = \phi_{n+1} f$$

Since $\phi_n = \phi_{n+1}$ on K_n we get

$$\frac{\partial(v - u_n)}{\partial \bar{z}} = 0 \ \text{and so} \ v - u_n \in Hol(Int\, K_n)$$

By the note we may find a $w \in Hol(\Omega)$ such that $|v - u_n - w| < 2^{-n}$ on K_{n-1}. We add $u_{n+1} := v - w$ to the system.

In this way we produce a sequence $u_1, u_2, \cdots$ from $C^\infty(\Omega)$ such that

$$\frac{\partial u_n}{\partial \bar{z}} = \phi_n f \ ,$$

$$u_{n+1} - u_n \in Hol(Int\, K_n) \ \text{and}$$

$$|u_{n+1} - u_n| < \frac{1}{2^n} \ \text{on } K_{n-1}$$

Any compact subset of Ω is contained in some K_N (by property (c) of Proposition 9), so the sum

$$\sum_{n=1}^{\infty} (u_{n+1} - u_n)$$

converges uniformly on any compact subset of Ω. Thus

$$u := \sum_{n=1}^{\infty} (u_{n+1} - u_n) + u_1$$

is a well defined function on Ω. For any fixed N we have $u_{n+1} - u_n \in Hol(Int\, K_N)$ whenever $n \geq N$, so the right hand side of

$$(*) \quad u - u_N = \sum_{n=N}^{\infty} (u_{n+1} - u_n)$$

is on $Int\, K_N$ the limit of a uniformly convergent sequence of holomorphic functions, so it is itself holomorphic on $Int\, K_N$ (Weierstrass' theorem). In particular $u \in C^\infty(Int\, K_N)$, and since N is arbitrary, $u \in C^\infty(\Omega)$.

When we apply $\partial/\partial\bar{z}$ to (*) on $Int\,K_N$ the right hand side vanishes (it is holomorphic), so

$$\frac{\partial(u - u_N)}{\partial\bar{z}} = 0 \ \text{ on } Int\,K_N$$

so

$$\frac{\partial u}{\partial\bar{z}} = \frac{\partial u_N}{\partial\bar{z}} = \phi_N f = f \ \text{ on } \ Int\,K_N$$

Since N is arbitrary we read that $\partial u/\partial\bar{z} = f$ throughout Ω. $\qquad\qquad$ $\square$

Section 5 Exercises

1. Generalize The Argument Principle as follows: Replace the condition about simple connectivity by the condition that γ is a closed path in Ω with the property that

$$Ind_\gamma(z) = 0 \ \text{ for all } \ z \in \mathbf{C}\backslash\Omega$$

2. Show that the polynomial $z \to z^4 + z^3 + 5z^2 + 2z + 4$ has no zeros in the first quadrant.

3. Let f be meromorphic in a neighborhood of $B[0, 1]$, and assume that there are no poles on the unit circle $|z| = 1$. Let $a \in \mathbf{C}$ satisfy

$$|a| > \sup_{|z|=1} |f(z)|$$

Show that f has the same number of poles and a-values (both counted with multiplicity) in $B(0, 1)$.

4. Consider the holomorphic function $f(z) := \sinh(\pi z)$ in the half strip $\{x + iy \mid x > 0, -\frac{1}{2} < y < \frac{1}{2}\}$. Show that f takes each value in the upper half plane $\{w \in \mathbf{C} \mid \Im z > 0\}$ exactly once.

5. Find the number of zeros of the polynomial $z \to z^7 - 5z^4 + z^2 - 2$ in the annulus $\{z \in \mathbf{C} \mid 1 < |z| < 2\}$.

6. Show that the polynomial

$$z \to 3z^{15} + 4z^8 + 6z^5 + 19z^4 + 3z + 1$$

has 4 zeros in $|z| < 1$ and 11 zeros in $1 < |z| < 2$.

7. Find a ball $B(0, R)$ such that the polynomial $z^3 - 4z^2 + z - 4$ has exactly two roots in it.

8. Let $a \in \mathbf{C}$, $|a| > e$ and let $n \in \{1, 2, 3, \cdots\}$. Show that the function $f(z) := e^z - az^n$ has exactly n zeros in $B(0, 1)$, and that they are different from one another.

9. Let $\lambda > 1$. Show that the equation $ze^{\lambda-z} = 1$ has exactly one root z in $B[0, 1]$, and that the root belongs to the open interval $]0,1[$.

10. Let $\lambda > 1$. Show that the equation $\lambda - z - e^{-z} = 0$ has exactly one root z_0 in the half plane $\{z \in \mathbf{C} \mid \Re z \geq 0\}$, and that $z_0 > 0$. What happens as $\lambda \to 1$?

11. Show that all four roots of the polynomial $z^4 - z + 5$ lie in the annulus $\{z \in \mathbf{C} \mid 1.35 < |z| < 1.65\}$, and that there is exactly one root in each open quadrant.

12. Consider the polynomial $P(z) = a_0 + a_1 z + \cdots + a_n z^n$ where $0 < a_0 < a_1 < \cdots < a_n$. Show that P has n zeros in $B(0,1)$. Hint: Consider $(1-z)P(z)$.

13. Use Rouché's theorem to prove that an nth order polynomial has exactly n roots.

14. Use Rouché's theorem to prove that the positions of the roots of an nth order polynomial depend continuously on the coefficients of the polynomial.

15. Prove the following lemma:

Lemma.

Suppose $a_1, a_2, \cdots, a_n$ are holomorphic functions on the disc $B(0,R)$, and suppose that $w_0 \in \mathbf{C}$ is a simple root of the polynomial

$$w \to w^n + a_1(0)w^{n-1} + \cdots + a_n(0)$$

Then there exist an $r \in]0, R[$, and a $\phi \in Hol(B(0,r))$ such that $\phi(0) = w_0$ and

$$(\phi(z))^n + a_1(z)(\phi(z))^{n-1} + \cdots + a_n(z) = 0 \ \text{for all} \ z \in B(0,r)$$

16. Let f be analytic on a neighborhood of $B[0,1]$ and assume that $f(B[0,1]) \subseteq B(0,1)$. Show that f has exactly one fixed point.

17. Prove the following theorem:

Theorem.

Let $f_1, f_2, \cdots$ be a sequence of functions which are holomorphic on an open set Ω. Assume that $f_n \to f_0$ locally uniformly on Ω as $n \to \infty$. Assume furthermore that f_0 has a zero of order n_0 at $z_0 \in \Omega$.

Then to every $r > 0$ there exists an $N = N(r)$ such that $n > N$ => the function f_n has at least n_0 zeros in the set $\Omega \cap B(z_0, r)$.

18. Let Ω be a connected open subset of $\mathbf{C}$. Assume that $f_n \in Hol(\Omega)$ are univalent for $n = 1, 2, \cdots$ and that $f_n \to f$ uniformly on compacta as $n \to \infty$.

Show that f is either univalent or constant.

19. Show that the open set $\mathbf{C} \backslash S$ is simply connected, where S is the infinite spiral (Archimedes' spiral)

$$S := \left\{ z = re^{ir} \mid 0 \leq r < \infty \right\}$$

20. Let Ω be an open connected non-empty subset of $\mathbf{C}$. Show that the following 8 statements are equivalent.

(α) Ω is simply connected.

(β) $Ind_\gamma(z) = 0$ for all closed curves γ in Ω and all $z \in \mathbf{C} \backslash \Omega$.

(γ) $\mathbf{C}_\infty \backslash \Omega$ is connected.

(δ) $\int_\gamma f = 0$ for each $f \in Hol(\Omega)$ and each closed path γ in Ω.

(ϵ) Every $f \in Hol(\Omega)$ has a primitive.

145

(ζ) Every never-vanishing $f \in Hol(\Omega)$ has a continuous logarithm.

(η) Every never-vanishing $f \in Hol(\Omega)$ has a continuous square root.

(θ) Ω is homeomorphic to $B(0,1)$.

21. Show the following theorem:

Theorem.

Let K be a compact subset of $\mathbf{C}$ with the property that $\mathbf{C} \backslash K$ is connected. Let f be holomorphic in an open neighborhood of K. Then f is on K a uniform limit of polynomials.

22. Let

$$\Omega := B(0,1) \backslash B\left[\frac{1}{2}, \frac{1}{2}\right]$$

and let $f \in Hol(\Omega)$.

(α) Does there exist a sequence of polynomials converging uniformly to f on Ω ?

(β) Given a compact subset K of Ω, does there always exist a sequence of polynomials which converges uniformly to f on K ?

(γ) Does there always exist a sequence of polynomials which converges uniformly to f on Ω, if f is even holomorphic on an open set containing the closure of Ω ?

23. Does there exist a sequence $\{P_n\}$ of polynomials such that for all $z \in \mathbf{C}$ we have as $n \to \infty$:

$$P_n(z) \to \begin{cases} 1 & \text{if } \Im z > 0 \\ 0 & \text{if } \Im z = 0 \\ 1 & \text{if } \Im z < 0 \end{cases}$$

Chapter 10 Representations of Meromorphic Functions

Infinite products and construction of meromorphic functions with prescribed zeros and poles are the main themes of this chapter. The Euler product for sine, the partial fraction expansion of cotangent and the Γ-function will be the showpieces.

Section 1 Infinite products

It is easy to construct a holomorphic function which has zeros exactly at finitely many given points $z_1, z_2, \cdots, z_N \in \mathbf{C}$. Simply take the function

$$z \rightarrow (z - z_1)(z - z_2) \cdots (z - z_N) = \prod_{k=1}^{N} (z - z_k)$$

With an infinity of points $z_1, z_2, \cdots$ one is tempted to take $\prod_{k=1}^{\infty} (z - z_k)$. Although this product does not converge, a suitable modification (where we insert suitable convergence factors) does, as we shall see in Weierstrass' factorization theorem. So we are led to study infinite products.

An infinite sum behaves in most respects like a finite one. Analogously an infinite product should behave like a finite product; in particular $\prod_{k=1}^{\infty} c_k = 0$ should mean that at least one of the factors c_k vanishes. However, examples like

$$\lim_{n \to \infty} \frac{1}{2}\frac{1}{2} \cdots \frac{1}{2} \, (n \text{ factors}) = 0$$

demonstrate that we must be cautious, at least if the limit is zero. That is the background for the following rather involved definition of an infinite product.

Definition 1.

Let $c_1, c_2, \cdots$ be a sequence of complex numbers. The infinite product

$$(1) \quad \prod_{k=1}^{\infty} c_k$$

is said to converge if there to each $\epsilon > 0$ exists an $N \in \mathbf{N}$ such that for all $n \geq N$ and $p \geq 1$:

$$(2) \quad \left| \prod_{k=n+1}^{n+p} c_k - 1 \right| < \epsilon$$

If the infinite product (1) converges, then the numbers $C_n := \prod_{k=1}^{n} c_k$, $n = 1, 2, \cdots$ form a bounded sequence. Using (2) we find that $\{C_n\}$ is a Cauchy sequence and therefore has a limit in $\mathbf{C}$. We use the notation $\prod\limits_{k=1}^{\infty} c_k$ also for this limit and say that it is the *value* of the infinite product (1).

If a finite number of factors is inserted in or removed from a convergent infinite product then the result will still be a convergent infinite product (and the value of the result will be the obvious one).

Taking $p = 1$ in (2) we see that $c_n \to 1$ as $n \to \infty$ if the product (1) converges. That is the reason why it is often convenient to write an infinite product in the form $\prod\limits_{k=1}^{\infty} (1 + a_k)$.

Also note that if $\prod\limits_{k=1}^{\infty} c_k$ is a convergent product, then $\prod\limits_{k=1}^{\infty} c_k = 0$ if and only if at least one of the factors c_k is zero.

We leave it to the reader to verify the following two quite elementary properties of infinite products:

$\prod\limits_{k=1}^{\infty} c_k$ and $\prod\limits_{k=1}^{\infty} d_k$ both convergent

$\Rightarrow \prod\limits_{k=1}^{\infty} c_k d_k$ converges and $\prod\limits_{k=1}^{\infty} c_k d_k = \left(\prod\limits_{k=1}^{\infty} c_k \right) \left(\prod\limits_{k=1}^{\infty} d_k \right)$.

$\prod\limits_{k=1}^{\infty} c_k$ convergent and $\prod\limits_{k=1}^{\infty} c_k \neq 0$

$$\Rightarrow \prod_{k=1}^{\infty} \frac{1}{c_k} \text{ converges and } \prod_{k=1}^{\infty} \frac{1}{c_k} = \frac{1}{\prod\limits_{k=1}^{\infty} c_k}$$

The following result turns out to be very useful, because it connects infinite products with sums.

Proposition 2.

The infinite product $\prod\limits_{k=1}^{\infty} (1 + a_k)$ converges if $\sum\limits_{k=1}^{\infty} |a_k| < \infty$, and in that case

$$(3) \quad \left| \prod_{k=1}^{\infty} (1 + a_k) - 1 \right| \leq \exp \left(\sum_{k=1}^{\infty} |a_k| \right) - 1$$

Proof : We first show the inequality

$$(4) \quad \left| \prod_{k=1}^{n} (1 + a_k) - 1 \right| \leq \prod_{k=1}^{n} (1 + |a_k|) - 1$$

The finite product minus 1, i.e. $\prod\limits_{k=1}^{n} (1 + a_k) - 1$ is a polynomial in $a_1, a_2, \cdots, a_n$ with positive coefficients, say

$$p(a_1, a_2, \cdots, a_n) = \prod_{k=1}^{n} (1 + a_k) - 1$$

p has positive coefficients so

$$|p(a_1, a_2, \cdots, a_n)| \le p(|a_1|, |a_2|, \cdots, |a_n|)$$

which is the inequality (4).

Noting that $1 + x \le e^x$ for any real x we put $x = |a_k|$ in (4) and deduce the finite version (3') of (3):

$$(3') \quad \left| \prod_{k=1}^{n} (1 + a_k) - 1 \right| \le \exp\left(\sum_{k=1}^{n} |a_k| \right) - 1$$

Coming now to the proof of the proposition we get from the inequality (3') applied to $a_{n+1}, a_{n+2}, \cdots, a_{n+p}$ that

$$\left| \prod_{k=n+1}^{n+p} (1 + a_k) - 1 \right| \le \exp\left(\sum_{k=n+1}^{n+p} |a_k| \right) - 1$$

If $\sum_{k=1}^{\infty} |a_k| < \infty$ then the right hand side is small, independent of $p \ge 1$ if only n is large. $\qquad\square$

We next turn to infinite products where the factors are analytic functions. First a natural definition.

Definition 3.

Let $a_1, a_2, \cdots$ be complex valued functions defined in an open subset Ω of $\mathbf{C}$. We say that the infinite product $\prod_{k=1}^{\infty} (1 + a_k)$ converges locally uniformly in Ω, if $\prod_{k=1}^{\infty} (1 + a_k(z))$ converges at each $z \in \Omega$ and if furthermore to each compact subset K of Ω and each $\epsilon > 0$ there exists an N such that for all $z \in K$ and $n \ge N$

$$\left| \prod_{k=1}^{\infty} (1 + a_k(z)) - \prod_{k=1}^{n} (1 + a_k(z)) \right| < \epsilon$$

Lemma 4.

Let $a_1, a_2, \cdots$ be holomorphic on an open subset Ω of $\mathbf{C}$.

(α) If the infinite product $\prod_{k=1}^{\infty} (1 + a_k)$ converges locally uniformly in Ω, then the limit function $F(z) := \prod_{k=1}^{\infty} (1 + a_k(z))$ is holomorphic on Ω.

F vanishes precisely at those points where at least one of the factors $1 + a_k$ vanishes. In particular F doesn't vanish identically on a connected open set unless at least one of the factors does so.

(β) If as $N \to \infty$ the sum $\sum_{n=N}^{\infty} |a_n(z)|$ converges locally uniformly to 0, then the infinite product $\prod_{k=1}^{\infty} (1 + a_k)$ converges locally uniformly in Ω.

Proof: The last statement (β) of the lemma is a consequence of Proposition 2 and the inequality (3). $\square$

It has already been observed that a convergent infinite product vanishes at a point iff at least one of its factors does so. If F vanishes identically on a connected open subset Ω_0 of Ω then at least one of the factors vanishes at uncountably many points of Ω_0. In that case the zeros of that factor has a limit point in Ω_0. By the Unique Continuation Theorem (Theorem IV.11) the factor is identically 0 on Ω_0.

Definition 5.

If F is holomorphic then F'/F is called the logarithmic derivative of F.

Placing ourselves in the situation of Lemma 4 where $F(z) = \prod_{k=1}^{\infty} (1 + a_k(z))$ we compute formally to get $\log F(z) = \sum_{k=1}^{\infty} \log (1 + a_k(z))$ which by formal differentiation yields

$$(5) \quad \frac{F'(z)}{F(z)} = \sum_{k=1}^{\infty} \frac{a_k'(z)}{1 + a_k(z)} \quad \text{for } z \in \Omega$$

Even though it would be tricky to justify this way of deriving the formula (5) for the logarithmic derivative of an infinite product, the formula is true, and it is possible to prove it directly. That is done in the next proposition.

Proposition 6.

Let $a_1, a_2, \cdots$ be holomorphic in an open subset Ω of $\mathbf{C}$ and suppose that the sum $\sum_{n=N}^{\infty} |a_n(z)|$ as $N \to \infty$ converges locally uniformly in Ω to 0. Let F denote the value of the infinite product $F(z) = \prod_{k=1}^{\infty} (1 + a_k(z))$ and put $\Omega_0 = \{z \in \Omega | F(z) \neq 0\}$.

Then

$$\sum_{k=1}^{\infty} \frac{a_k'(z)}{1 + a_k(z)}$$

converges absolutely for any $z \in \Omega_0$, and it converges locally uniformly on Ω_0 towards the meromorphic function F'/F, so

$$(5) \quad \frac{F'(z)}{F(z)} = \sum_{k=1}^{\infty} \frac{a_k'(z)}{1 + a_k(z)} \quad \text{for } z \in \Omega_0$$

Proof: The finite product $F_N(z) = \prod_{k=1}^{N} (1 + a_k(z))$ converges as $N \to \infty$ locally uniformly on Ω to F (Lemma 4). By Weierstrass' Theorem (Theorem IV.10) the sequence of derivatives $\{F_N'\}$ converges locally uniformly on Ω to F'.

Let K be a compact subset of Ω_0. Then F_N'/F_N converges uniformly on K to F'/F. Since

$$\frac{F_N'(z)}{F_N(z)} = \sum_{k=1}^{N} \frac{a_k'(z)}{1 + a_k(z)} \quad \text{for} \ z \in \Omega_0$$

we have only left to show the claim on absolute convergence, i.e. that

$$\sum_{k=1}^{\infty} \frac{|a_k'(z_0)|}{|1 + a_k(z_0)|}$$

converges for each $z_0 \in \Omega_0$.

Given $z_0 \in \Omega_0$ there exists an N such that $|a_k(z_0)| \leq \frac{1}{2}$ for $k \geq N$, and hence that the denominator $|1 + a_k(z_0)| \geq \frac{1}{2}$ for $k \geq N$. Thus it suffices to verify that $\sum_{k=1}^{\infty} |a_k'(z_0)| < \infty$. For that purpose choose complex numbers ϵ_k for $k = 1, 2, \cdots$ such that $|\epsilon_k| = 1$ and $\epsilon_k a_k'(z_0) = |a_k'(z_0)|$. By assumption the series $\sum_{k=1}^{\infty} \epsilon_k a_k$ converges locally uniformly on Ω, hence (Weierstrass' Theorem) so does its derivative $\sum_{k=1}^{\infty} \epsilon_k a_k'$. In particular it converges at $z = z_0$, which is the desired statement. $\square$

Section 2 The Euler formula for sine

The infinite product

$$F(z) := z \prod_{n=1}^{\infty} \left(1 - \frac{z^2}{n^2}\right)$$

converges (by Lemma 4) locally uniformly in the entire z-plane, so that F is an entire function. The zero-set of F is $\mathbf{Z}$ and each zero has multiplicity one. The function $z \to \sin \pi z$ has the same properties. Hence $z \to \sin(\pi z)/F(z)$ is holomorphic and zero free in the entire z-plane. We can write

$$(6) \quad \sin \pi z = F(z) e^{A(z)}$$

where A is entire. The main difficulty is the determination of A. To find A we take logarithmic derivatives in (6) :

$$(7) \quad \pi \cot \pi z = A'(z) + \frac{1}{z} + \sum_{n=1}^{\infty} \left(\frac{1}{z+n} + \frac{1}{z-n}\right)$$

The series on the right hand of (7) converges locally uniformly on the open set $\mathbb{C}\backslash\mathbf{Z}$. Differentiating (7) we get

$$(8) \quad \frac{-\pi^2}{\sin^2\pi z} = A''(z) - \frac{1}{z^2} - \sum_{n=1}^{\infty}\left(\frac{1}{(z+n)^2} + \frac{1}{(z-n)^2}\right)$$

and since $\sum_{n=1}^{\infty} n^{-2}$ converges, we may rewrite (8) in the form

$$(9) \quad A''(z) = \sum_{n=-\infty}^{\infty} \frac{1}{(z+n)^2} - \frac{\pi^2}{\sin^2\pi z}$$

The right hand side in (9) is unaltered if we replace z by $z+1$. In other words, the entire function A'' has period 1. To show that A'' is a constant it is thus by Liouville's theorem sufficient to verify that A'' is bounded in the set $\{z = x + iy \,|\, 0 \le x \le 1, |y| \ge 1\}$. For such z we have

$$\left|\sum_{n=-\infty}^{\infty}\frac{1}{(z+n)^2}\right| \le \sum_{n=-\infty}^{\infty}\frac{1}{|z+n|^2} = \sum_{n=0}^{\infty}\frac{1}{(x+n)^2+y^2} + \sum_{n=-\infty}^{-1}\frac{1}{(x+n)^2+y^2}$$

$$= \sum_{n=0}^{\infty}\frac{1}{(x+n)^2+y^2} + \sum_{n=1}^{\infty}\frac{1}{(n-x)^2+y^2}$$

$$= \sum_{n=0}^{\infty}\frac{1}{(x+n)^2+y^2} + \sum_{n=0}^{\infty}\frac{1}{(n+(1-x))^2+y^2}$$

Since $0 \le x \le 1$ we estimate as follows

$$(x+n)^2 \ge n^2 \quad \text{and} \quad (n+(1-x))^2 \ge n^2$$

and get

$$\left|\sum_{n=-\infty}^{\infty}\frac{1}{(z+n)^2}\right| \le \sum_{n=0}^{\infty}\frac{1}{n^2+y^2} + \sum_{n=0}^{\infty}\frac{1}{n^2+y^2} = 2\sum_{n=0}^{\infty}\frac{1}{n^2+y^2}$$

If $y > N$ (N positive integer) then

$$2\sum_{n=0}^{\infty}\frac{1}{n^2+y^2} \le 2\sum_{n=0}^{\infty}\frac{1}{n^2+N^2} \le 2\sum_{n=0}^{\infty}\frac{1}{\frac{1}{2}(n+N)^2} = 4\sum_{n=N}^{\infty}\frac{1}{n^2}$$

For the other term of the right hand side of (9) we estimate as follows for $y > 0$ (The case $y < 0$ can be treated similarly):

$$|\sin\pi z|^2 = \left|\frac{e^{\pi iz} - e^{-\pi iz}}{2}\right|^2 = \left|\frac{e^{\pi ix}e^{-\pi y} - e^{-\pi ix}e^{\pi y}}{2}\right|^2$$

$$= \left|\frac{e^{\pi y} - e^{2\pi ix}e^{-\pi y}}{2}\right|^2 \ge \left(\frac{e^{\pi y} - e^{-\pi y}}{2}\right)^2 = (\sinh\pi y)^2$$

so for any y we have

$$\left| \frac{\pi^2}{\sin^2 \pi z} \right| \le \frac{\pi^2}{\sinh^2 \pi y}$$

These inequalities say that $A''(z)$ tends to 0 as z tends to infinity in the period strip $0 \le \Re z \le 1$, so we conclude that $A'' = 0$, i.e. that A' is a constant. By (7) A' is odd, so $A' = 0$, i.e. A is a constant.

Finally, dividing both sides of (6) by z and letting z tend to 0 we evaluate the constant e^A to be π, and arrive at the beautiful *Euler formula for sine* :

$$(10) \quad \sin \pi z = \pi z \prod_{n=1}^{\infty} \left(1 - \frac{z^2}{n^2} \right)$$

Let us note that we on the way also found (formula (7)) the *partial fraction expansion of cotangent* :

$$(11) \quad \pi \cot \pi z = \frac{1}{z} + \sum_{n=1}^{\infty} \left(\frac{1}{z+n} + \frac{1}{z-n} \right) \quad \text{for } z \in \mathbf{C} \backslash \mathbf{Z}$$

As an application of (10) take $z = 1/2$ to get

$$1 = \frac{\pi}{2} \prod_{n=1}^{\infty} \left(1 - \frac{1}{4n^2} \right) = \frac{\pi}{2} \prod_{n=1}^{\infty} \frac{4n^2 - 1}{4n^2}$$

from which we obtain *Wallis' product* :

$$\frac{\pi}{2} = \prod_{n=1}^{\infty} \frac{4n^2}{4n^2 - 1} = \prod_{n=1}^{\infty} \frac{(2n)^2}{(2n - a)(2n + 1)} = \frac{2 \cdot 2}{1 \cdot 3} \cdot \frac{4 \cdot 4}{3 \cdot 5} \cdot \frac{6 \cdot 6}{5 \cdot 7} \cdots$$

As another illustration divide both sides of (10) by $1 - z$ and let $z \to 1$ to get

$$\frac{1}{2} = \prod_{n=2}^{\infty} \left(1 - \frac{1}{n^2} \right)$$

For another treatment of the topics of this section see [Wal].

Section 3 Weierstrass' factorization theorem

In contrast to the example of the last section consider the problem of constructing an entire function with simple zeros exactly at the points $0, -1, -2, -3, \cdots$ The product $z \prod_{n=1}^{\infty} \left(1 + \frac{z}{n} \right)$ will not do, simply because it does not converge. To treat such cases Weierstrass introduced certain explicitly given factors - the so-called primary factors - into the product to force convergence. Before we take up the general case let us illustrate the idea by the following lemma:

Lemma 7.

Let $a_1, a_2, \cdots$ be different complex numbers with the property that

$$\sum_{n=1}^{\infty} \frac{1}{|a_n|^2} < \infty$$

Then the infinite product

$$(12) \quad \prod_{k=1}^{\infty} \left\{ \left(1 - \frac{z}{a_k}\right) e^{\frac{z}{a_k}} \right\}$$

converges locally uniformly on $\mathbf{C}$ and its value is an entire function with simple zeros precisely at a_k, $k = 1, 2, 3, \cdots$

Proof : We start with an estimate, valid for all $|a| < 1$:

$$(13) \quad |(1-a)e^a - 1| \le |a|^2$$

To prove (13) note that we for $|a| < 1$ have

$$|(1-a)e^a - 1| = \left| \sum_{n=2}^{\infty} a^n \left(\frac{1}{n!} - \frac{1}{(n-1)!} \right) \right|$$

$$\le |a|^2 \sum_{n=2}^{\infty} |a|^{n-2} \left(\frac{1}{(n-1)!} - \frac{1}{n!} \right)$$

$$\le |a|^2 \sum_{n=2}^{\infty} \left(\frac{1}{(n-1)!} - \frac{1}{n!} \right) = |a|^2$$

Now, $\sum_{1}^{\infty} |a_n|^{-2} < \infty$ so $|a_n|$ must tend to infinity as $n \to \infty$. For z in a fixed compact set K we have $|z/a_n| < 1$ uniformly in z for all large n. Using (13) we realize that the series

$$\sum_{k=N}^{\infty} \left| \left(1 - \frac{z}{a_k}\right) e^{\frac{z}{a_k}} - 1 \right|$$

as $N \to \infty$ converges to 0 locally uniformly on $\mathbf{C}$. Lemma 4 now takes over. $\qquad \square$

In the general case we will modify the infinite product $\prod_{1}^{\infty} (1 - z/a_n)$ by replacing each factor $(1 - z/a_n)$ by a factor of the form

$$(14) \quad \left(1 - \frac{z}{a_n}\right) \exp \lambda\left(\frac{z}{a_n}\right)$$

We choose the functions λ in such a way that the new factors are so close to 1 that the product converges. That $(1 - z) \exp(\lambda(z))$ is close to 1 means that $\lambda(z)$ approximates $\log\left((1-z)^{-1}\right)$. The technical details are as follows:

Definition 8.

The primary factors of Weierstrass *are the entire functions*

$$E(z, 0) = 1 - z$$

$$E(z, n) = (1 - z) \exp\left(z + \frac{z^2}{2} + \cdots + \frac{z^n}{n}\right) \ for \ n = 1, 2, \cdots$$

Note that the primary factors vanish only at $z = 1$.

We will need the estimate

$$(15) \quad |E(z, n) - 1| \le |z|^{n+1} \text{ for } n = 0, 1, \cdots \text{ and } |z| \le 1$$

which shows that $E(z, n)$ is close to 1 for large n, even though it is zero at $z = 1$.

To prove (15) we find by a simple calculation that

$$-E'(z, n) = z^n \exp\left(z + \frac{z^2}{2} + \cdots + \frac{z^n}{n}\right)$$

so the power series expansion of $-E'(z, n)$ around $z = 0$ starts from z^n and all its coefficients are nonnegative. From

$$1 - E(z, n) = \int\limits_{[0, z]} \left(-E'(w, n)\right) dw$$

we then see that the power series expansion of $1 - E(z, n)$ starts from z^{n+1} and has nonnegative coefficients as well, i.e.

$$1 - E(z, n) = \sum_{k=n+1}^{\infty} a_k z^k \text{ where all } a_k \ge 0$$

So for $|z| \le 1$ we get

$$|1 - E(z, n)| \le \sum_{k=n+1}^{\infty} a_k |z|^k \le |z|^{n+1} \sum_{k=n+1}^{\infty} a_k 1 = |z|^{n+1}(1 - E(1, n))$$

$$= |z|^{n+1}(1 - 0) = |z|^{n+1}$$

which is (15).

And then for any $m \ge 0$, $n \ge 0$ and $|z| \le 1$:

$$(16) \quad |E(z, n)^m - 1| \le 2^m |E(z, n) - 1| \le 2^m |z|^{n+1}$$

where we have used the identity

$$|a^m - 1| = |a - 1| |a^{m-1} + a^{m-2} + \cdots + a + 1|$$

and the fact (evident from (16)) that $|E(z, n)| \le 2$ for $|z| \le 1$.

In particular if $n = m + k$ then for any $m \geq 0$ and $k \geq 0$:

$$(17) \quad |E(z, m + k) - 1| \leq 2^{-k} \text{ for } |z| \leq \frac{1}{2}$$

With these preliminaries to a side we can prove *Weierstrass' factorization theorem*:

Theorem 9 (Weierstrass' factorization theorem).

Let Ω be an open subset of $\mathbf{C}$, let $z_1, z_2, \cdots$ be a sequence of different points from Ω with no cluster point in Ω, and let finally $m_1, m_2, \cdots$ be a sequence of positive integers.

There exists a function $g \in Hol(\Omega)$ such that g has a zero of order m_k at z_k for $k = 1, 2, \cdots$, and such that g vanishes nowhere else in Ω.

Proof :

Since the sequence $\{z_k\}$ does not cluster in Ω, there exist $a \in \Omega$ and $r > 0$ such that $B(a, r) \subseteq \Omega$ and such that $B(a, r)$ does not intersect the set $\{z_k\}$. Let us for convenience assume that $a = 0$ and $r = 1$. Under the transformation $z \to z^{-1}$ the open set $\Omega \backslash \{0\}$ goes into an open set V. If $c_k := z_k^{-1}$, then $|c_k| \leq 1$, $c_k \in V$ and $\{c_k\}$ does not cluster in V (Recall that 0 does not belong to V). Choose for each k a point $a_k \in \mathbf{C} \backslash V$ closest to c_k. Since $\{c_k\}$ does not cluster in V we get

$$(18) \quad \lim_{k \to \infty} |a_k - c_k| = 0$$

[Indeed, if for some $\delta > 0$ and subsequence $\{k_i\}$ we have that $|a_{k_i} - c_{k_i}| \geq \delta$, where c_{k_i} tends to c (recall that $\{c_k\}$ is bounded), then c must belong to $\mathbf{C} \backslash V$. But by choice of a_k we have

$$|c_{k_i} - c| \geq |c_{k_i} - a_{k_i}| \geq \delta$$

which is a contradiction].

Now put

$$(19) \quad f(z) := \prod_{k=1}^{\infty} E\left(\frac{c_k - a_k}{z - a_k}, m_k + k\right)^{m_k} \text{ for } z \in V$$

If F is any compact subset of V, then by (18) for all large k, $2|c_k - a_k| < |z - a_k|$, uniformly for $z \in F$.

The local uniform convergence of the infinite product in (19) now follows from (17) and Lemma 4.

By assumption $B(0, 1) \subseteq \Omega$, so $\{z \,|\, |z| > 1\} \subseteq V$. Since $|a_k| \leq 1$ and $|c_k| \leq 1$, we have

$$\left|\frac{c_k - a_k}{z - a_k}\right| \leq \frac{1}{2} \text{ whenever } |z| \geq 3$$

Using (17) and (19) we see that

$$(20) \quad |f(z)| \text{ is bounded for } |z| \geq 3$$

Now define $g(z) := f(1/z)$ for $z \neq 0$. Then g has a zero of order m_k at z_k and vanishes nowhere else in $\Omega \backslash \{0\}$. (20) says that g is bounded in $B(0, \frac{1}{3}) \backslash \{0\}$. 0 is therefore a removable singularity of g and g can hence be defined to be analytic at 0 as well. If $g(0) \neq 0$, g is the required function. If 0 is a zero of g of order n, then $z^{-n}g$ fulfills all the requirements. $\qquad \square$

Section 4 The Γ-function

Taking $a_n = -n$ in (12) we find that

$$(21) \quad g(z) := e^{\gamma z} z \prod_{n=1}^{\infty} \left(1 + \frac{z}{n}\right) e^{-\frac{z}{n}}$$

is entire with simple zeros precisely at $z = 0, -1, -2, \cdots$. Here γ is a constant chosen so that $g(1) = 1$. γ is called the *Euler constant*. The reciprocal of g is meromorphic in **C** with simple poles at $0, -1, -2, \cdots$ This is the famous Γ-*function* (gamma-function):

$$(22) \quad \frac{1}{\Gamma(z)} = g(z)$$

and it is of utmost importance. The Γ-function was introduced by Euler in 1729 (See [Ds] for a readable account of its history).

Let us derive some of the most important properties of the Γ-function. Since g is entire, Γ has no zeros. Writing

$$(23) \quad g(z) = e^{\gamma z} z(z+1) e^{-z} \prod_{n=1}^{\infty} \left\{ \left(1 + \frac{z}{n+1}\right) \exp\left(-\frac{z}{n+1}\right) \right\}$$

and replacing z by $z + 1$ in (21), the following computation is valid for any $z \neq 0, -1, -2, \cdots$

$$\frac{g(z)}{zg(z+1)} = \frac{e^{-z}}{e^{\gamma}} \prod_{n=1}^{\infty} \left\{ \frac{1 + \frac{z}{n+1} \exp\left(-\frac{z}{n+1}\right)}{1 + \frac{z+1}{n} \exp\left(-\frac{z+1}{n}\right)} \right\}$$

$$= e^{-z-\gamma} \prod_{n=1}^{\infty} \left\{ \frac{n}{n+1} \exp\left(\frac{z}{n(n+1)} + \frac{1}{n}\right) \right\}$$

$$= e^{-z-\gamma} \exp\left\{ z \sum_{n=1}^{\infty} \frac{1}{n(n+1)} \right\} \prod_{n=1}^{\infty} \left\{ \frac{n}{n+1} \exp\left(\frac{1}{n}\right) \right\}$$

Using that

$$\sum_{n=1}^{\infty} \frac{1}{n(n+1)} = \sum_{n=1}^{\infty} \left(\frac{1}{n} - \frac{1}{n+1}\right) = 1$$

we find further that

$$(24) \quad \frac{g(z)}{zg(z+1)} = e^{-\gamma} \prod_{n=1}^{\infty} \left\{ \frac{n}{n+1} \exp\left(\frac{1}{n}\right) \right\}$$

Thus $g(z) = czg(z)$ for $z \neq 0, -1, -2, \cdots$, where c is a constant. By continuity this holds for all $z \in \mathbf{C}$. To find the constant c we note from (21) that

$$\lim_{z \to 0} \frac{g(z)}{z} = 1$$

Since also $g(1) = 1$ we see that the constant is 1. Translating from g to Γ we find the *functional equation for* Γ :

$$(25) \ \Gamma(z+1) = z\Gamma(z) \ \text{for} \ z \neq 0, -1, -2, \cdots$$

Since $\Gamma(1) = 1$, (25) leads to

$$(26) \ \Gamma(n+1) = n! \ \text{for} \ n = 0, 1, 2, \cdots$$

which explains why the Γ-function is also called the *factorial function*.

We proceed by deriving another fundamental relation for the Γ-function, called the formula of complementary arguments. Use (21) with z and $-z$ and multiply to get

$$g(z)g(-z) = -z^2 \prod_{n=1}^{\infty} \left(1 - \frac{z^2}{n^2}\right)$$

Appealing to the Euler formula (10) for sine and to (25) the above relation reads in terms of Γ :

$$(27) \ \Gamma(z)\Gamma(1-z) = \frac{\pi}{\sin \pi z} \ \text{for} \ z \neq \mathbf{Z}$$

(27) is the *formula of complementary arguments*.

In particular since $\Gamma(x) > 0$ for $x > 0$, we get taking $z = 1/2$ in (27) that

$$(28) \ \Gamma\left(\frac{1}{2}\right) = \sqrt{\pi}$$

The reader might in other connections have encountered the Mellin transform (= the Euler integral) of a complex valued function g on $\mathbf{R}^+$, which is the function Γ_g defined by

$$\Gamma_g(x) := \int_0^{\infty} t^{x-1} g(t) dt \ \text{wherever the integral converges.}$$

The particular case of $g(t) = e^{-t}$ is of special interest for us here because this Γ_g equals the Γ-function. This result is called *The Euler integral formula*

$$(29) \ \Gamma(z) = \int_0^{\infty} t^{z-1} e^{-t} dt \ \text{for} \ \Re z > 0$$

158

To derive it we use Prym's decomposition

$$(30) \quad \int\limits_0^\infty t^{z-1}e^{-t}dt = \int\limits_0^1 t^{z-1}e^{-t}dt + \int\limits_1^\infty e^{-t}t^{z-1}dt = P(z) + Q(z)$$

The first integral on the right, i.e. $Q(z)$ is entire. Expanding e^{-t} and integrating term by term we find that the second integral equals

$$(31) \quad P(z) = \sum_{n=0}^\infty \frac{(-1)^n}{n!(z+n)}$$

The right hand side of (31) converges uniformly in any compact set not containing the points $z = 0, -1, -2, \cdots$, and we see that

$$(32) \quad F(z) := \sum_{n=0}^\infty \frac{(-1)^n}{n!(z+n)} + Q(z)$$

is a meromorphic function in the entire complex plane with simple poles at $z = 0, -1, -2, \cdots$. It also equals the integral in (29) for $\Re z > 0$. Using this, routine integration shows that

$$(33) \quad zF(z) = F(z+1) \quad \text{when} \quad \Re z > 0$$

But then (33) must hold in the entire z-plane minus the points $0, -1, -2, \cdots$ (The Unique Continuation Theorem). We know that Γ satisfies the identity (33) in form of (25) and that Γ never vanishes. Thus from (22) we see that

$$(34) \quad G := \frac{F}{\Gamma} = Fg$$

is entire and has period 1, i.e. $G(z+1) = G(z)$. Since $F(1) = \Gamma(1) = 1$, $G - 1$ vanishes at the (real) integers. Now, $\sin \pi z$ has period 2, has simple poles at the integers and vanishes nowhere else. Comparing G with $\sin \pi z$ we see that

$$(35) \quad H(z) := \frac{G(z) - 1}{\sin \pi z}$$

is entire and has period 2. We will prove that H is zero by showing that H is bounded in the period strip $1 \le \Re z \le 3$, and then applying Liouville's theorem.

F is bounded in this strip, as is evident from its integral representation (30). Let us estimate g in the strip. If $z = x + iy$ is in the strip under consideration then

$$\left| \left(1 + \frac{z}{n}\right)e^{-\frac{z}{n}} \right| = \sqrt{\left(1 + \frac{x}{n}\right)^2 + \frac{y^2}{n^2}} e^{-\frac{x}{n}} \le \left(1 + \frac{x}{n}\right)e^{-\frac{x}{n}}\sqrt{1 + \frac{y^2}{n^2}}$$

Using this estimate in (21) and separating products we get:

$$|g(z)| \le |z|e^{\gamma x}\left\{ \prod_{n=1}^\infty \left(1 + \frac{x}{n}\right)e^{-\frac{x}{n}} \right\}\sqrt{\prod_{n=1}^\infty \left(1 + \frac{y^2}{n^2}\right)}$$

$$= |z|\frac{g(x)}{x}\sqrt{\prod_{n=1}^\infty \left(1 + \frac{y^2}{n^2}\right)} = |z|\frac{g(x)}{x}\sqrt{\frac{\sin \pi i y}{\pi i y}}$$

where the last equality comes from the Euler sine formula (10). Using this estimate it is elementary to show that (remember that $g(x)$ is bounded for $1 \leq x \leq 3$ and that $\sin(\pi i y) = i \sinh(\pi y)$)

$$(36) \quad \frac{g(z)}{\sin \pi z} \to 0 \text{ as } z \to \infty \text{ in } 1 \leq \Re z \leq 3$$

From (34) and (35)

$$|H(z)| \leq |F(z)| \left| \frac{g(z)}{\sin \pi z} \right| + \frac{1}{|\sin \pi z|}$$

Using (36) we conclude that H is bounded in the period strip and hence everywhere. By Liouville's theorem H is a constant. Since $H(z)$ tends to 0 as z tends to ∞ in the period strip, the constant is 0, so H is identically 0, i.e. $F = \Gamma$. We have thus derived (29). $\qquad\square$

From (28) we then get by a change of variables the important formula

$$(37) \quad \int\limits_{-\infty}^{\infty} e^{-x^2} dx = \int\limits_{0}^{\infty} t^{-\frac{1}{2}} e^{-t} dt = \sqrt{\pi}$$

As an application of the Euler integral formula (29) we shall derive Stirling's asymptotic approximation of the factorial function:

Theorem 10 (Stirling's approximation formula).

$$\frac{\Gamma(n) e^n \sqrt{n}}{n^n} \to \sqrt{2\pi} \text{ as } n \to \infty$$

Proof: The proof is adapted from the one in [Pa].
Introducing $x = \sqrt{t} - \sqrt{n}$ as new variable in (29) we find

$$\frac{\Gamma(n) e^n \sqrt{n}}{n^n} = 2 \int\limits_{-\sqrt{n}}^{\infty} \left(1 + \frac{x}{\sqrt{n}} \right)^{2n+1} e^{-2\sqrt{n}x} e^{-x^2} dx$$

Since $1 + y \leq \exp y$ we get for $x > -\sqrt{n}$ the following estimate of the integrand (except for the factor $\exp(-x^2)$)

$$\left(1 + \frac{x}{\sqrt{n}} \right)^{2n-1} e^{-2\sqrt{n}x} \leq \exp \left\{ \frac{x}{\sqrt{n}}(2n-1) \right\} \exp(-2\sqrt{n}x)$$

$$= \exp \left(-\frac{x}{\sqrt{n}} \right) \leq e$$

so the integrand above is bounded uniformly in n by the integrable function $e \exp(-x^2)$.

We next show that the integrand converges pointwise, so x is fixed for a moment. Using the power series expansion of the logarithm we find for $n \to \infty$ that

$$Log\left(1 + \frac{x}{\sqrt{n}}\right) = \frac{x}{\sqrt{n}} - \frac{x^2}{2n} + O\left(\frac{1}{n^{\frac{3}{2}}}\right)$$

and so

$$\log\left\{\left(1 + \frac{x}{\sqrt{n}}\right)^{2n-1} e^{-2\sqrt{n}x}\right\} = (2n-1)\log\left(1 + \frac{x}{\sqrt{n}}\right) - 2\sqrt{n}x$$

$$= (2n-1)\left(\frac{x}{\sqrt{n}} - \frac{x^2}{2n}\right) + O\left(\frac{1}{\sqrt{n}}\right) - 2\sqrt{n}x = -x^2 + O\left(\frac{1}{\sqrt{n}}\right) \to -x^2$$

Finally, by the dominated convergence theorem

$$\frac{\Gamma(n)e^n\sqrt{n}}{n^n} \to 2\int_{-\infty}^{\infty} e^{-x^2}e^{-x^2}dx = \sqrt{2\pi} \text{ as } n \to \infty$$

$\square$

As the end of this paragraph we mention the famous *duplication formula of Legendre* (see the exercises for a proof):

$$(38) \quad \sqrt{\pi}\Gamma(2z) = 2^{2z-1}\Gamma(z)\Gamma\left(z + \frac{1}{2}\right)$$

Section 5 The Mittag-Leffler expansion

The theorem of this paragraph is analogous to the Weierstrass theorem on zero sets in §4. Instead of zeros we deal with poles. Before we state and prove the theorem we remind the reader of the following fact from the theory of Laurent expansions:

Theorem 11.

Let f be holomorphic on the punctured disc $B(0,R)\backslash\{0\}$. Then f may in exactly one way be written in the form

$$f(z) = g(z) + h\left(\frac{1}{z}\right) \text{ for } z \in B(0,R)\backslash\{0\}$$

where $g \in Hol(B(0,R))$ and where h is an entire function such that $h(0) = 0$.

Proof: This was seen during the proof of Theorem VI.3. $\square$

More generally, if $a \in \mathbb{C}$ is an isolated singularity of a holomorphic function f then f can be written in the form

$$(39) \quad f(z) = h\left(\frac{1}{z-a}\right) + g(z)$$

where h is an entire function with $h(0) = 0$, and where g is a holomorphic function on the domain of definition of f and a is a removable singularity for g. The decomposition (39) is unique. The first term, i.e. the function $z \to h\left((z-a)^{-1}\right)$ is called the *principal part of f at a.*

Example.

The function (Cf formula (9) above)

$$\frac{\pi^2}{\sin^2(\pi z)} = \sum_{n=-\infty}^{\infty} \frac{1}{(z-n)^2}$$

has for each $n \in \mathbf{N}$ the principal part $(z-n)^{-2}$ at $z = n$.

However, if we are given a sequence of principal parts, then their sum will normally not converge. Mittag-Leffler's theorem says that a meromorphic function with the prescribed principal parts nevertheless exists:

Theorem 12 (Mittag-Leffler's theorem (1884)).

Let Ω be an open subset of $\mathbf{C}$. Let $p_1, p_2, \cdots$ be a sequence of different points from Ω without cluster point in Ω, and let $P_1, P_2, \cdots$ be a sequence of polynomials without constant term.

Then there exists a function which is meromorphic on Ω, has $\{p_1, p_2, \cdots\}$ as its set of poles in Ω and whose principal part at p_k is $P_k\left((z-p_k)^{-1}\right)$ for all $k = 1, 2, \cdots$

Remark. If we only specify that the desired function should have poles at $p_1, p_2, \cdots$ of given orders $m_1, m_2, \cdots$, then it can be found by help of Weierstrass' theorem on zero sets from §4: Indeed, if f is a holomorphic function with zeros at $p_1, p_2, \cdots$ of orders $m_1, m_2, \cdots$ then $1/f$ satisfies the requirements.

Proof :

Let us write $f_k(z) = P_k\left((z-p_k)^{-1}\right)$ for $k = 1, 2, \cdots$. Let $B(p_k, r_k)$, $k = 1, 2, \cdots$ be disjoint balls in Ω and choose for each k a function $\phi_k \in C_0^\infty(B(p_k, r_k))$ such that ϕ_k is identically 1 on the smaller ball $B_k := B(p_k, r_k/2)$.

The function $f := \sum_{k=1}^{\infty} \phi_k f_k$ is smooth on $\Omega \backslash \{p_1, p_2, \cdots\}$ and reduces on B_k to f_k, so it satisfies the requirements of the theorem except that it is not holomorphic. To remedy that we consider the function

$$g(z) := \begin{cases} \frac{\partial f}{\partial \bar{z}}(z) & \text{for } z \in \Omega \backslash \{p_1, p_2, \cdots\} \\ 0 & \text{for } z \in \{p_1, p_2, \cdots\} \end{cases}$$

g is clearly C^∞ on $\Omega \backslash \{p_1, p_2, \cdots\}$. And on the punctured disc $B_k \backslash \{p_k\}$ the function f reduces to a holomorphic function (viz f_k), so

$$g = \frac{\partial f}{\partial \bar{z}} = \frac{\partial f_k}{\partial \bar{z}} = 0$$

on $B_k\backslash\{p_k\}$. Thus $g \in C^\infty(\Omega)$.

The inhomogeneous Cauchy-Riemann equation $\partial u/\partial\bar{z} = g$ has a solution $u \in C^\infty(\Omega)$ (Theorem IX.16). Note that $\partial u/\partial\bar{z} = g = 0$ on B_k, so that $u \in Hol(B_k)$.

Now, $h := f - u$ is holomorphic in $\Omega\backslash\{p_1, p_2, \cdots\}$ by the very definition of g. On $B_k\backslash\{p_k\}$ we have $h = f_k - u$. As observed $u \in Hol(B_k)$, so h has the desired principal part f_k at p_k. □

Weierstrass' factorization theorem (Theorem 9) told us how to construct a holomorphic function whose zero set is a prescribed sequence of points. The next result is that what can be done with zeros can be done with any sequence of values:

Corollary 13 (Germay's interpolation theorem).

Let Ω be an open subset of the complex plane. Let $\{z_1, z_2, \cdots\}$ be a subset of Ω without limit points in Ω, and let $w_1, w_2, \cdots$ be a sequence of complex numbers.

Then there exists a function $f \in Hol(\Omega)$ such that $f(z_k) = w_k$ for $k = 1, 2, \cdots$

Proof :

By Weierstrass' Factorization theorem (Theorem 9) there is a holomorphic function f_0 with simple poles at the z_k, i.e. $f_0(z_k) = 0$ and $f_0'(z_k) \neq 0$ for $k = 1, 2, \cdots$ By the Mittag-Leffler theorem there is a function $h \in Hol(\Omega\backslash\{z_1, z_2, \cdots\})$ such that

$$z \to h(z) - \frac{w_k}{f_0'(z_k)}\frac{1}{z - z_k}$$

is holomorphic in a ball around z_k for each k. Now, $f := f_0 h$ is holomorphic on Ω, because the zeros cancel the singularities, and for $z \to z_k$ we find that

$$f(z) = f_0(z)h(z) = f_0(z)\left\{h(z) - \frac{w_k}{f_0'(z_k)}\frac{1}{z - z_k}\right\} + f_0(z)\frac{w_k}{f_0'(z_k)}\frac{1}{z - z_k}$$
$$\to 0 + w_k = w_k$$

so $f(z_k) = w_k$ for $k = 1, 2, \cdots$ □

For a generalization of Corollary 13 see Exercise 17.

Section 6 The ζ- and $\wp$-functions of Weierstrass

We have examples of periodic holomorphic functions. E.g. exp, cos and sin. Liouville's theorem rules out that a nonconstant holomorphic function can have two independent periods. We shall now give an example of a doubly periodic meromorphic function, viz Weierstrass' $\wp$-function.

Let ω and ω' be non-zero complex numbers with non-real quotient, i.e. they are linearly independent over the reals. Let G be the group generated by ω and ω', i.e. G consists of the complex numbers of the form $g = m\omega + n\omega'$ where m and n are integers. If ω and ω' are periods of a given function then so is every $g \in G$.

By Mittag-Leffler's theorem there is a meromorphic function with principal part $(z-g)^{-1}$ at each element $g \in G$. However we do not have to appeal to that theorem, since an example is given by the so-called *Weierstrass ζ-function* (Weierstrass zeta-function):

$$(42) \quad \zeta(z) := \frac{1}{z} + \sum_{0 \neq g \in G} \left\{ \frac{1}{z-g} + \frac{1}{g} + \frac{z}{g^2} \right\}$$

To prove that ζ is such an example we prepare a lemma.

Lemma 14.

$$(43) \quad \sum_{0 \neq g \in G} \frac{1}{|g|^3} < \infty$$

Proof: Each $g \in G$, $g \neq 0$ has the form $g = m\omega + n\omega'$ where $|m| + |n| \geq 1$. If $|m| + |n| = k$ either $|m|$ or $|n|$ is $\geq k/2$; if, say $|m| \geq k/2$ then

$$\left| m\omega + n\omega' \right| = |\omega'| \left| m\frac{\omega}{\omega'} + n \right| \geq |\omega'| \left| \Im \frac{m\omega}{\omega'} \right| = |\omega'| |m| \left| \Im \frac{\omega}{\omega'} \right| \geq k\alpha$$

where $\alpha = 2\min\left\{ |\omega'| \left| \Im \frac{\omega}{\omega'} \right|, |\omega| \left| \Im \frac{\omega'}{\omega} \right| \right\}$.

Now there are at most $4k$ pairs (m,n) such that $|m| + |n| = k$. Thus

$$\sum_{0 \neq g \in G} \frac{1}{|g|^3} \leq 4 \sum_{k=1}^{\infty} \sum_{|m|+|n|=k} \frac{1}{\alpha^3 k^3} \leq 4\alpha^{-3} \sum_{k=1}^{\infty} \frac{1}{k^2} < \infty$$

which proves (43). □

Let us now return to the right hand side of (42). Consider the disc $B(0,R)$. There are only finitely many $g \in G$ satisfying $|g| \leq R$, as is clear from f.ex.. (43). Since

$$\frac{1}{z-g} + \frac{1}{g} + \frac{z}{g^2} = \frac{z^2}{(z-g)g^2} \quad , \text{ and}$$

$$|z-g| = |g| \left| 1 - \frac{z}{g} \right| \geq \frac{|g|}{2} \text{ if } |g| \geq 2R \text{ and } |z| \leq R$$

the series

$$\sum_{|g| \geq 2R} \left\{ \frac{1}{z-g} + \frac{1}{g} + \frac{z}{g^2} \right\}$$

converges uniformly in $B(0,R)$ and represents a holomorphic function in this disc. It follows that $\zeta(z)$, given by (42) is meromorphic in $\mathbb{C}$ with simple poles precisely at the points of G.

Differentiation of (42) termwise (which is permitted) leads to the *Weierstrass $\wp$-function* (Weierstrass pe-function):

$$(44) \quad \wp(z) = -\zeta'(z) = \frac{1}{z^2} + \sum_{0 \neq g \in G} \left\{ \frac{1}{(z-g)^2} - \frac{1}{g^2} \right\}$$

The $\wp$-function is meromorphic in $\mathbf{C}$ with double poles precisely at $g \in G$. Let us now show that the $\wp$-function is doubly periodic with periods ω and ω'. Differentiating (44) we get

$$(45) \quad \wp'(z) = -\frac{2}{z^3} - \sum_{0 \neq g \in G} \frac{2}{(z-g)^3} = -\sum_{g \in G} \frac{2}{(z-g)^3}$$

For any $h \in G$ the series for $\wp'$ is unaltered by a change from z to $z + h$ because G is a group. In other words, $\wp'(z + h) = \wp'(z)$, implying that

$$(46) \quad \wp(z+h) - \wp(z) = C(h)$$

where $C(h)$ is a constant, perhaps depending on h. But the series for $\wp$ shows that $\wp$ is an even function : $\wp(z) = \wp(-z)$. Now it is immediate from (46) that

$$C(h + g) = C(h) + C(g) \quad \text{for all} \ \ h, g \in G$$

Using these facts we find that

$$C(-h) = \wp(z - h) - \wp(z) = \wp(-z - h) - \wp(-z) = \wp(z + h) - \wp(z) = C(h)$$

But $C(h) + C(-h) = C(0) = 0$, so we must have $C(h) = 0$ and so $\wp(z + h) = \wp(z)$. We record this in the final statement of this section:

$\wp$ and $\wp'$ are doubly periodic with periods ω and ω', and $\zeta' = -\wp$.

Section 7 Exercises

1. Show that

$$\prod_{n=1}^{\infty} \left(1 - \frac{2}{(n+1)(n+2)} \right) = \frac{1}{3}$$

2. For which $z \in \mathbf{C}$ will the infinite product

$$\prod_{n=0}^{\infty} \left(1 + z^{(2^n)} \right)$$

converge? Show that the value of the product is $(1 - z)^{-1}$.

Hint:

$$(1 - z) \prod_{n=1}^{k-1} \left(1 + z^{(2^n)} \right) = 1 - z^{(2^k)}$$

3. Assume that the limit $\lim\limits_{N\to\infty} \prod\limits_{n=1}^{N} c_n$ exists and is $\neq 0$. Show that the product $\prod\limits_{n=1}^{\infty} c_n$ converges.

4. Assume that the infinite product $\prod\limits_{n=1}^{\infty} (1 + a_n)$ converges. Discuss convergence of $\prod\limits_{n=1}^{\infty} \sqrt{1 + a_n}$.

5. Let the sequence $\{a_n\}$ of complex numbers satisfy that

$$0 < |a_n| < 1 \quad \text{for} \quad n = 1, 2, \cdots \text{ and that } \sum_{n=1}^{\infty} (1 - |a_n|) < \infty.$$

(i) Show that the infinite product (a so-called *Blaschke product*)

$$B(z) := \prod_{n=1}^{\infty} \frac{|a_n|}{a_n} \frac{a_n - z}{1 - \overline{a_n} z}$$

defines a function which is holomorphic in $B(0, 1)$. Find its zeros.

(ii) Find a sequence $\{a_n\}$ as above with the property that every point on the unit circle $|z| = 1$ is a cluster point of $\{a_n\}$.

6. We shall in this exercise present another way of dealing with the entire function A from §2 (Herglotz' trick):

$$A'(z) = \pi \cot(\pi z) - \left(\frac{1}{z} + \sum_{n=1}^{\infty} \frac{2z}{z^2 - n^2} \right)$$

(α) Show that A' satisfies the functional equation

$$A'(z) = \frac{1}{2} \left\{ A'\left(\frac{z}{2} \right) + A'\left(\frac{z + 1}{2} \right) \right\}$$

(β) Use the functional equation to show that A' is a constant. (Hint: The maximum modulus principle).

7. Show that

$$\sum_{n=1}^{\infty} \frac{1}{n^2} = \frac{\pi^2}{6}$$

by help of the partial fraction expansion of the cotangent.

8. Show that

$$\cos \pi z = \prod_{n=1}^{\infty} \left\{ 1 - \frac{4z^2}{(2n - 1)^2} \right\}$$

9. Let f be meromorphic on an open subset Ω of **C**. Show that f can be written as a quotient $f = g/h$, where g and h both are holomorphic in Ω.

10. Let $t \in \mathbf{R} \backslash \{0\}$. Show that

$$|\Gamma(it)| = \sqrt{\frac{\pi^2}{t \sinh(\pi t)}}$$

11. (i) Show that Euler's constant equals

$$\gamma = \lim_{N \to \infty} \left(1 + \frac{1}{2} + \cdots + \frac{1}{N} - \log N \right)$$

(ii) Derive *Gauss' formula*

$$\Gamma(z) = \lim_{N \to \infty} \frac{N! N^z}{z(z+1)\cdots(z+N)} \quad \text{for } z \neq 0, -1, -2, \cdots$$

and from it deduce Wallis' product

$$\frac{\pi}{2} = \frac{2 \cdot 2}{1 \cdot 3} \cdot \frac{4 \cdot 4}{3 \cdot 5} \cdot \frac{6 \cdot 6}{5 \cdot 7} \cdots$$

(iii) Prove the *Legendre duplication formula*

$$\sqrt{\pi}\Gamma(2z) = 2^{2z-1}\Gamma(z)\Gamma\left(z + \frac{1}{2}\right)$$

12. Find the residue of Γ at $z = 0$. More generally at $z = -n$ for $n = 0, 1, 2, \cdots$.

13. Let C be the following contour

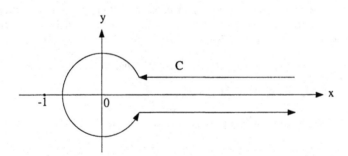

Prove *Hankel's formula*

$$\Gamma(z) = \frac{i}{2\sin(\pi z)} \int_C (-t)^{z-1} e^{-t} dt$$

For which $z \in \mathbf{C}$ is Hankel's formula valid?

14. Let Ω and $B = \{b_1, b_2, \cdots\}$ be as in the Mittag-Leffler theorem. Let f_k for $k = 1, 2, \cdots$ be entire functions with $f_k(0) = 0$.

Show that there exists an $f \in Hol(\Omega \backslash B)$ such that the principal part of f at b_k is $f_k\left((z - b_k)^{-1}\right)$ for all $k = 1, 2, \cdots$

15. With the notation of §6 show that

$$\sigma(z) := z \prod_{0 \neq g \in G} \left\{ \left(1 - \frac{z}{g}\right) \exp\left(\frac{z}{g} + \frac{1}{2}\left(\frac{z}{g}\right)^2\right) \right\}$$

is entire with simple poles exactly at the $g \in G$.

The logarithmic derivative of σ is Weierstrass' ζ-function. σ is called the *sigma-function of Weierstrass*.

16. Show the following generalization of Proposition 2 :

Let $\mu_1, \mu_2, \cdots$ be a sequence of reals with $\mu_k > 0$ for all k. Let $c_1, c_2, \cdots$ be a sequence of complex numbers from the open unit disc, and assume that

$$\sum_{k=1}^{\infty} |c_k| \mu_k < \infty$$

Show that the product

$$\prod_{k=1}^{\infty} (1 + c_k)^{\mu_k}$$

converges (principal determinations).

17. Generalize Corollary 13 to the following theorem:

Theorem.

Let Ω be an open subset of the complex plane and let $\{z_1, z_2, \cdots\}$ be a subset of Ω without limit points in Ω. Let there for each $k = 1, 2, \cdots$ be given finitely many complex numbers $w_{k,1}, w_{k,2}, \cdots, w_{k,N_k}$. Then there exists a function $f \in Hol(\Omega)$ such that

$$f^{(n)}(z_k) = w_{k,n} \text{ for } 0 \leq n \leq N_k \text{ and } k = 1, 2, \cdots$$

Hint: Modify the proof of Corollary 13.

18. Derive the formula (37) directly without resort to the Γ—function as follows:

Let

$$I := \int_{-\infty}^{\infty} e^{-x^2} dx \ ,$$

show that

$$I^2 = \int_{\mathbf{R}^2} e^{-\left(x^2 + y^2\right)} dx dy$$

and introduce polar coordinates.

Another elegant way of proving (37) can be found in [An].

Chapter 11 The Prime Number Theorem

Section 1 The Riemann zeta function

The famous prime number theorem states that $\pi(x)$, the number of primes less than or equal to x, is asymptotically the same as $x/\log x$. This was conjectured by Gauss and about a hundred years later proved by Hadamard and de la Vallée-Poussin. It is this proof that we present below. An "elementary" proof was given by A. Selberg [Se].

The proof uses the celebrated *Riemann zeta function* defined for z in the open half plane $\{w \in \mathbf{C} \mid \Re w > 1\}$ by:

$$(1) \quad \zeta(z) := \sum_{n=1}^{\infty} \frac{1}{n^z}$$

The ζ-function was studied already by Euler, long before the time of Riemann (See [Ay]). Euler found e.g. formulas for $\zeta(2n)$, $n = 1, 2, \cdots$. These formulas are derived in an elementary fashion in [Ber]. The special case $\zeta(2) = \pi^2/6$ has a short elegant proof [Ap2]. Very little is known about the values of the ζ-function at other natural numbers, except that it has been established that $\zeta(3)$ is irrational. See the informal report [Po].

We shall presently show that the function given by (1) for $\Re z > 1$ is the restriction to this open set of a meromorphic function on $\mathbf{C}$ with a simple pole at $z = 1$. It is the extended function that is usually called the ζ-function. It is remarkable that the properties of this function should relate so decisively to the prime number theorem. For more information on the ζ-function we refer to the monograph [Ed].

Theorem 1.

The series (1) converges uniformly on every half plane of the form $\{z \in \mathbf{C} \mid \Re z \geq a\}$ where $a > 1$.

There is a meromorphic function ζ on $\mathbf{C}$ such that $\zeta(z) - (z-1)^{-1}$ is entire and such that $\zeta(z)$ is given by (1) for $\Re z > 1$.

Proof: We concentrate on the second statement and leave the first one to the reader. For $\Re z > 1$ we have

$$\Gamma(z) = \int_0^{\infty} t^{z-1} e^{-t} dt = n^z \int_0^{\infty} t^{z-1} e^{-nt} dt$$

so therefore

$$(2) \quad \zeta(z)\Gamma(z) = \sum_{n=1}^{\infty} \int_0^{\infty} t^{z-1} e^{-nt} dt = \int_0^{\infty} \frac{t^{z-1}}{e^t - 1} dt$$

169

where interchange of sum and integral is justified by the dominated convergence theorem. Write

$$(3) \quad \zeta(z)\Gamma(z) = \left(\int_0^1 + \int_1^\infty \right) \frac{t^{z-1}}{e^t - 1} dt$$

Noting that the second integral is entire in z, we concentrate on the first one. The function $z/(e^z - 1)$ is holomorphic in the ball $|z| < 2\pi$, so we may expand it

$$\frac{z}{e^z - 1} = \sum_{n=0}^\infty a_n z^n \quad \text{for } |z| < 2\pi$$

The series converges in particular for $z = 2$, so $a_n 2^n \to 0$. Thus there is a constant C such that $|a_n| \leq C 2^{-n}$ for $n = 0, 1, 2, \cdots$. Since $a_0 = 1$, we have

$$\frac{1}{e^z - 1} = \frac{1}{z} + \sum_{n=1}^\infty a_n z^{n-1}$$

Using this expansion and term by term integration (justified by uniform convergence) the first integral in (3) can be written

$$(5) \quad \int_0^1 \frac{t^{z-1}}{e^t - 1} dt = \frac{1}{z-1} + \sum_{n=1}^\infty \frac{a_n}{z + n - 1}$$

Thus we may for $\Re z > 1$ write

$$(6) \quad \zeta(z)\Gamma(z) = \frac{1}{z-1} + \sum_{n=1}^\infty \frac{a_n}{z + n - 1} + \int_1^\infty \frac{t^{z-1}}{e^t - 1} dt$$

As said before the third term on the right in (6) is entire; the second term is by our estimate on a_n meromorphic in the complex plane with simple poles at $-(n-1)$ if $a_n \neq 0$, $n = 1, 2, \cdots$. And the first term has a simple pole at $z = 1$. Writing $g(z) = 1/\Gamma(z)$ we may define ζ on the complex plane by

$$(7) \quad \zeta(z) = \frac{g(z)}{z-1} + \left(\sum_{n=1}^\infty \frac{a_n}{z + n - 1} \right) g(z) + g(z) \int_1^\infty \frac{t^{z-1}}{e^t - 1} dt$$

Since g is entire with zeros at $0, -1, -2, \cdots$ the right hand side of (7) is meromorphic on all of $\mathbf{C}$ with at most one pole, namely one at $z = 1$. Since $g(1) = 1$, the function $\zeta(z) - 1/(z-1)$ is entire. $\qquad \square$

Riemann discovered a remarkable functional relation for the ζ-function from which many of its properties can be deduced.

Theorem 2 (The functional equation for the zeta-function).
The following relation is valid:

$$(8) \quad \zeta(1-z) = \frac{2}{(2\pi)^z} \zeta(z)\Gamma(z) \cos\frac{\pi z}{2}$$

or written in a more symmetric form

$$(8') \quad \zeta(z)\Gamma\left(\frac{z}{2}\right)\pi^{-\frac{z}{2}} = \zeta(1-z)\Gamma\left(\frac{1-z}{2}\right)\pi^{-\frac{1-z}{2}}$$

Proof : Integrate by parts in (2) for $\Re z > 1$ to obtain

$$\zeta(z)\Gamma(z) = \frac{1}{z}\int_0^\infty \frac{t^z e^t}{(e^t-1)^2}dt = -\frac{1}{4\pi}\int_0^\infty \frac{t^z}{\sin^2\left(\frac{it}{2}\right)}$$

so that

$$(9) \quad z\zeta(z)\Gamma(z) = -\int_0^1 \frac{t^z}{4\sin^2\left(\frac{it}{2}\right)}dt - \int_1^\infty \frac{t^z}{4\sin^2\left(\frac{it}{2}\right)}dt$$

$$= -\int_0^1 t^z\left\{\frac{1}{4\sin^2\left(\frac{it}{2}\right)} + \frac{1}{t^2}\right\}dt + \frac{1}{z-1} - \int_1^\infty \frac{t^z}{4\sin^2\left(\frac{it}{2}\right)}dt$$

Now from the formula (8) of Chapter X we get

$$(10) \quad \frac{1}{4\sin^2\left(\frac{it}{2}\right)} = -\frac{1}{t^2} + \sum_{n=1}^\infty \left\{\frac{1}{(it+2\pi n)^2} + \frac{1}{(it-2\pi n)^2}\right\}$$

The function $1 - \frac{\sin z}{z}$ behaves like $z^2/3!$ near $z = 0$, so

$$1 - \frac{\sin^2 z}{z^2} = \left(1 + \frac{\sin z}{z}\right)\left(1 - \frac{\sin z}{z}\right)$$

behaves like $z^2/3$ near $z = 0$. Hence the first integral in (9) is holomorphic for $\Re z > -1$. The last integral is in fact entire. Thus (9) holds for $\Re z > -1$. However, for $\Re z < 1$

$$\frac{1}{z-1} = -\int_1^\infty t^{z-2}dt$$

so we can certainly for $-1 < \Re z < 1$ write

$$(11) \quad z\zeta(z)\Gamma(z) = -\int_0^\infty t^z\left\{\frac{1}{4\sin^2\left(\frac{it}{2}\right)} + \frac{1}{t^2}\right\}dt$$

171

And now we can use the expansion in (10). We may interchange the sum and the integral because of dominated convergence: Indeed, if $-1 < \Re z < 0$, then

$$\int\limits_0^\infty t^{\Re z} \sum_{n=1}^\infty \frac{1}{|2\pi n + it|^2} dt = \sum_{n=1}^\infty \int\limits_0^\infty \frac{t^{\Re z}}{4\pi^2 n^2 + t^2} dt$$

$$= \sum_{n=1}^\infty \int\limits_0^\infty \frac{(2\pi n)^{\Re z} s^{\Re z}}{4\pi^2 n^2 (1+s^2)} ds = \sum_{n=1}^\infty \frac{(2\pi n)^{\Re z+1}}{4\pi^2 n^2} \int\limits_0^\infty \frac{s^{\Re z}}{1+s^2} ds < \infty$$

Thus using (10) in (11), changing the summation and integration around and using the substitution $t = 2\pi n u$ we get

$$(12) \quad z\zeta(z)\Gamma(z) = -\sum_{n=1}^\infty (2\pi n)^{z-1} \int\limits_0^\infty \left\{ \frac{1}{(1+iu)^2} + \frac{1}{(1-iu)^2} \right\} u^z du$$

The sum on the right hand side is just $(2\pi)^{z-1}\zeta(1-z)$. The integral - which is independent of n - can be evaluated as follows: The integrand is

$$2\frac{1-u^2}{(1+u^2)^2} u^z$$

so the integral equals

$$2\int\limits_0^\infty \frac{1-u^2}{(1+u^2)^2} u^z du = 2\int\limits_0^\infty \frac{u^z}{1+u^2} du - 4\int\limits_0^\infty \frac{u^{2+z}}{(1+u^2)^2} du = -2z\int\limits_0^\infty \frac{u^z}{1+u^2} du$$

where we have just integrated by parts in the second integral. The integral on the right hand side is for $-1 < \Re z < 1$ evaluated below to be

$$(13) \quad 2\int\limits_0^\infty \frac{u^z}{1+u^2} du = \frac{\pi}{\cos\left(\frac{\pi z}{2}\right)}$$

Using (13) in (12) we conclude that (8) is valid in the strip $-1 < \Re z < 0$. Since both sides are meromorphic in all of the complex plane the relation is an identity. $\quad\square$

Proof of (13):

The substitution $v = u^2$ converts the above integral to

$$\int\limits_0^\infty \frac{v^{\frac{z-1}{2}}}{1+v} dv = \int\limits_0^\infty \frac{v^{\frac{z+1}{2}-1}}{1+v} dv$$

and the required result is a particular case of the formula

$$\int\limits_0^\infty \frac{x^{p-1}}{(1+x)^{p+q}} dx = \frac{\Gamma(p)\Gamma(q)}{\Gamma(p+q)} \quad \text{when } \Re p > 0, \Re q > 0$$

combined with the formula of complementary arguments X.(28). To derive the formula we write

$$\Gamma(p+q)\frac{x^{p-1}}{(1+x)^{p+q}} = \int_0^\infty t^{p+q-1}e^{-t}\frac{x^{p-1}}{(1+x)^{p+q}}dt$$

$$= \int_0^\infty e^{-t(1+x)}t^{p+q-1}x^{p-1}dt$$

Now we integrate both sides with respect to x from 0 to ∞ and change the order of integration on the right hand side (which is justified by absolute convergence):

$$\Gamma(p+q)\int_0^\infty \frac{x^{p-1}}{(1+x)^{p+q}}dx = \int_0^\infty e^{-t}t^{p+q-1}\left(\int_0^\infty e^{-tx}x^{p-1}dx\right)dt$$

$$= \int_0^\infty t^{q-1}e^{-t}\left(\int_0^\infty e^{-y}y^{p-1}dy\right)dt = \Gamma(p)\Gamma(q)$$

$\square$

Section 2 Euler's product formula and zeros of ζ

Euler's product formula relates the ζ-function directly to the prime numbers. Simple consequences are some results on the zeros of the ζ-function.

Theorem 3 (Euler's product formula).

$$(14)\quad \zeta(z) = \prod_p \frac{1}{(1-p^{-z})} \quad \text{when } \Re z > 1$$

where p in the infinite product ranges over all primes $p > 1$.

Proof: Let us first prove that the infinite product converges: If we for each prime p write

$$\frac{1}{1-p^{-z}} = 1 + \sum_{m=1}^\infty p^{-mz}$$

then it suffices by Proposition X.2 to show that

$$\sum_p \left|p^{-z} + p^{-2z} + p^{-3z} + \cdots\right| < \infty$$

Now

$$\sum_p \left|p^{-z} + p^{-2z} + p^{-3z} + \cdots\right| \le \sum_p \sum_{n=1}^\infty \left|p^{-nz}\right| \le \sum_p \sum_{n=1}^\infty \left(\frac{1}{p^n}\right)^{\Re z}$$

173

The right hand side is a subset of the terms from the series

$$\sum_{m=1}^{\infty} \left(\frac{1}{m}\right)^{\Re z}$$

so

$$\sum_{p} \left| p^{-z} + p^{-2z} + p^{-3z} + \cdots \right| \leq \sum_{m=1}^{\infty} \left(\frac{1}{m}\right)^{\Re z} = \zeta(\Re z) < \infty$$

To prove that the value of the infinite product actually is $\sum_{n=1}^{\infty} n^{-z}$ we consider the finite product

$$P(N) := \prod_{p \leq N} \left\{ 1 + p^{-z} + p^{-2z} + \cdots \right\} \quad \text{for } N \in \mathbf{N}$$

We can multiply this product out and rearrange the terms as we please without altering the result because it is just a product of (finitely many) absolutely convergent series. A typical term is of the form

$$p_1^{-\alpha_1 z} p_2^{-\alpha_2 z} \cdots p_k^{-\alpha_k z}$$

where $p_1, p_2, \cdots, p_k$ are different primes and $\alpha_1, \alpha_2, \cdots, \alpha_k$ are non-negative integers. By the fundamental theorem of arithmetic $P(N) = \sum_{n \in A} n^{-z}$, where $A = A(N)$ consists of those $n \in \mathbf{N}$ whose prime factors all are $\leq N$. Thus

$$\left| \sum_{n=1}^{\infty} \frac{1}{n^z} - P(N) \right| \leq \sum_{n \in B} \frac{1}{|n^z|}$$

where B consists of those n having at least one prime factor $> N$. In particular

$$\left| \sum_{n=1}^{\infty} \frac{1}{n^z} - P(N) \right| \leq \sum_{n > N} \frac{1}{|n^z|} = \sum_{n > N} \frac{1}{n^{\Re z}}$$

The right hand side converges to 0 as $N \to \infty$, so

$$P(N) \underset{N \to \infty}{\to} \sum_{n=1}^{\infty} \frac{1}{n^z}$$

as desired. $\qquad\qquad\qquad\qquad\qquad\qquad\qquad\qquad\qquad\qquad\qquad\qquad\qquad\quad$ □

As an amusing commentary on (14) we may note that there are infinitely many primes (that fact was already known and proved by Euclid): Indeed, if there were but finitely many then the ζ-function would not diverge at $z = 1$. Not only are there infinitely many primes, but as observed by Euler (1737) there are so many that the series formed by the reciprocals of the prime numbers diverges, i.e.

$$\sum_{p} \frac{1}{p} = \frac{1}{2} + \frac{1}{3} + \frac{1}{5} + \frac{1}{7} + \frac{1}{11} + \cdots = \infty$$

(See Exercise 1). For elementary derivations of this fact consult [AG] or [Ap1;Theorem 1.13].

Corollary 4.

The ζ-function does not vanish for $\Re z > 1$. For $\Re z < 0$ it vanishes only at the points $-2m$, $m = 1, 2, \cdots$.

Proof : The first assertion is immediate from (14). Since the Γ-function has no zeros, the second assertion follows from the first and the functional equation (8). $\quad\square$

Taking logarithmic derivatives in (14) we obtain for $\Re z > 1$ that

$$(15) \quad -\frac{\zeta'(z)}{\zeta(z)} = \sum_p \frac{Log\, p}{p^z} \frac{1}{1 - p^{-z}} = \sum_p Log\, p \sum_{n=1}^{\infty} (p^n)^{-z} = \sum_{m=2}^{\infty} \Lambda(m)m^{-z}$$

where

$$\Lambda(m) = \begin{cases} Log\, p & \text{if } m \text{ is a power } (> 0) \text{ of } p \\ 0 & \text{otherwise} \end{cases}$$

It is permitted to rearrange terms because of the absolute convergence (Cf Proposition X.6).

The fundamental link between the ζ-function and the distribution of primes is provided by the following function (called Chebyshev's function)

$$(16) \quad \psi(x) := \sum_{m \leq x} \Lambda(m) \text{ for } x \in [1, \infty[$$

The main result, known as the *Hadamard-de la Vallée Poussin theorem* asserts that

$$(17) \quad \lim_{x \to \infty} \frac{\psi(x)}{x} = 1$$

We will first show that the prime number theorem is a consequence of (17):

Given a prime p the number of k such that $p^k \leq x$ is less than or equal to $(Log\, x)/(Log\, p)$, so recalling that $\Lambda(m) = Log\, p$ if m is a power of p we get from the definition (16) that

$$(18) \quad \psi(x) \leq \sum_{p \leq x} \frac{Log\, x}{Log\, p} Log\, p = \pi(x) Log\, x$$

where $\pi(x) = $ the number of primes less than or equal to x.

On the other hand, if $1 < y < x$ then

$$(19) \quad \pi(x) - \pi(y) = \sum_{y < p \leq x} 1 \leq \sum_{y < p \leq x} \frac{Log\, p}{Log\, y} \leq \frac{\psi(x)}{Log\, y}$$

Taking $y = x/(Log\,x)^2$ and noting that $\pi(y) \leq y$ we obtain from (19) for any $x > e$ that

$$(20) \quad \pi(x)\frac{\log x}{x} \leq \frac{1}{\log x} + \frac{\psi(x)}{x}\frac{\log x}{\log x - 2\log\log x}$$

That the prime number theorem is equivalent to (17) is now evident from (18) and (20). $\qquad\square$

Before we embark on the proof of (17) we need one more fact, namely that ζ has no zeros on the line $\Re z = 1$. That result is due to Hadamard and de la Vallée Poussin.

Theorem 5.
The ζ-function has no zeros on the line $\Re z = 1$.

Proof: ζ is meromorphic, hence so is $\eta := \zeta'/\zeta$. Thus, for any complex number w, $\lim_{z \to w}(z - w)\eta(z)$ exists and is an integer. It is positive when w is a zero of ζ, negative when w is a pole of ζ, and zero otherwise.

If $z = 1 + \epsilon + it$ where $\epsilon > 0$ and $t \in \mathbf{R}$ we have from (15) that

$$\Re\{\eta(1 + \epsilon + it)\} = -\sum_{m=2}^{\infty} \Lambda(m)m^{-1-\epsilon}\cos(t\log m)$$

Using

$$0 \leq (1 + \cos\theta)^2 = 1 + 2\cos\theta + \cos^2\theta = \frac{1}{2}(3 + 4\cos\theta + \cos 2\theta)$$

we then get

$$(21) \quad 3\Re(\eta(1 + \epsilon)) + 4\Re(\eta(1 + \epsilon + it)) + \Re(\eta(1 + \epsilon + 2it))$$
$$= -\sum_{m=2}^{\infty} \Lambda(m)m^{-1-\epsilon}[1 + \cos(t\log m)]^2 \leq 0$$

Now multiply the above by ϵ and let ϵ tend to 0. If $1 + it$ were a zero of ζ then we get that the limit of the left hand side of (21) is $8 - 3 + 4 + 0 = 1$.

This contradiction proves Theorem 5. $\qquad\square$

Section 3 More about the zeros of ζ

In this section, which is a digression we mention couple of results on the zeros of ζ. We have seen that ζ has no zeros in the closed half plane $\Re z \geq 1$. The functional equation (8) then tells us that the only zeros in the half plane $\Re z \leq 0$ are $-2m$, $m = 1, 2, \cdots$. These latter zeros are called the trivial zeros. All the non-trivial zeros - if any - are thus in the open strip $0 < \Re z < 1$. This is called the *critical strip*.

Let us show that ζ has no *real* zeros in the critical strip: We start with the formula (2) or rather the consequence

$$\zeta(z)\Gamma(z) = \int\limits_0^1 \frac{t^{z-1}}{e^t-1}dt + \int\limits_1^\infty \frac{t^{z-1}}{e^t-1}dt$$

which we for $\Re z > 1$ rewrite as

$$\zeta(z)\Gamma(z) = \int\limits_0^1 t^{z-1}\left(\frac{1}{e^t-1}-\frac{1}{t}\right)dt + \frac{1}{z-1} + \int\limits_1^\infty \frac{t^{z-1}}{e^t-1}dt$$

The above formula is derived for $\Re z > 1$, but since the right hand side is meromorphic in $\Re z > 0$ it is valid in $\Re z > 0$. For $0 < t$ we have

$$\frac{1}{t} > \frac{1}{e^t-1}$$

so the first integral above is negative when $z > 0$. For $t \geq 1$ we have $e^t - 1 \geq t^2$ so the third term is at most

$$\int\limits_1^\infty \frac{dt}{t^2} = 1$$

Finally, for $0 < z < 1$ we have $1/(z-1) < -1$, so $\zeta(z)\Gamma(z) < 0$. $\qquad\square$

One of the most famous still unsolved problems in mathematics is the

Riemann hypothesis :
All non-trivial zeros of ζ lie on the line $\Re z = \frac{1}{2}$.

Numerical computations support the validity of the Riemann hypothesis. In [LR] it is shown that the ζ-function has exactly 300 000 001 zeros whose imaginary parts lie between 0 and 119 590 809.282 and all of them have real part 1/2. See also [Wag]. On the theoretical side we will here just mention that ζ has infinitely many zeros on the line $\Re z = \frac{1}{2}$, a result due to Hardy (1914).

Section 4 The prime number theorem

We now set out to prove the prime number theorem. It is not easy to explain why or how the method came about.

Let us start with the right hand side of (15). Using Abel's partial summation formula (Proposition I.10) we get

$$\sum_{m=2}^\infty \frac{\Lambda(m)}{m^z} = \sum_{m=2}^{N-1}\left\{\frac{1}{m^z}-\frac{1}{(m+1)^z}\right\}\psi(m) + \frac{\psi(N)}{N^z}$$

where ψ is defined in (16). Since $\pi(x) \le x$ the last term above tends to 0 as $N \to \infty$ by (18) when $\Re z > 1$. Thus

$$\sum_{m=2}^{\infty} \frac{\Lambda(m)}{m^z} = \sum_{m=2}^{\infty} \left\{ \frac{1}{m^z} - \frac{1}{(m+1)^z} \right\} \psi(m) = \sum_{m=1}^{\infty} \left\{ \frac{1}{m^z} - \frac{1}{(m+1)^z} \right\} \psi(m)$$

Since ψ(t) equals ψ(m) for $m \le t < m+1$ the last sum can be written

$$\sum_{m=1}^{\infty} z \int_{m}^{m+1} t^{-z-1} dt \psi(m) = z \int_{1}^{\infty} t^{-z-1} \psi(t) dt = z \int_{0}^{\infty} e^{-(z-1)y} \left\{ e^{-y} \psi(e^y) \right\} dy$$

where we have just introduced $y = \log t$ as new variable - it is much more convenient to deal with the interval $[0, \infty[$. Thus from (15) we get

$$(22) \quad -\frac{1}{z} \frac{\zeta'(z)}{\zeta(z)} = \int_{0}^{\infty} e^{-(z-1)t} H(t) dt \quad \text{when } \Re z > 1$$

where $H(t) := e^{-t} \psi(e^t)$.

Note that (17) is equivalent to $\lim_{t \to \infty} H(t) = 1$.

We can further rewrite (22) as

$$(23) \quad A(z) := -\frac{1}{z} \frac{\zeta'(z)}{\zeta(z)} - \frac{1}{z-1} = \int_{0}^{\infty} e^{-(z-1)t} (H(t) - 1) dt$$

The reason for doing so is: We know that $\zeta(z) - (z-1)^{-1}$ is entire (that is Theorem 1). A simple computation then shows that the function A is analytic wherever $z\zeta(z) \neq 0$; in particular (by Theorem 5) in a neighborhood of the line $\Re z = 1$.

The right hand side of (23) is the Laplace transform of the function $H-1$. Referring back to the fact that (17) is equivalent to $H(t) - 1 \to 0$ as $t \to \infty$, we now realize that we need to prove a type of Tauberian result (A Tauberian result derives the asymptotic behavior of a function from the behavior of its averages).

We define $\rho : \mathbf{R} \to \mathbf{R}$ by

$$\rho(y) := \begin{cases} \frac{1-|y|/2}{2} & \text{when } |y| \le 2 \\ 0 & \text{when } |y| > 2 \end{cases}$$

and let $u > 0$, $\lambda > 0$ be real parameters that later on in the proof will converge to ∞.

With $z = 1 + \epsilon + i\lambda y$, where $\epsilon > 0$, we multiply both sides of (23) by $\rho(y)e^{iuy}$ and integrate with respect to y to get

$$(24) \quad \int_{-2}^{2} A(1 + \epsilon + i\lambda y)\rho(y)e^{iuy} dy = \int_{-2}^{2} \rho(y)e^{iuy} \left\{ \int_{0}^{\infty} (H(t) - 1)e^{-\epsilon t - i\lambda yt} dt \right\} dy$$

Interchange of order of integration is permitted on the right hand side of (24) because

$$|H(t) - 1| \leq 1 + H(t) = 1 + e^{-t}\psi(e^t) \leq 1 + t$$

by (18), so the double integral is absolutely convergent. Effecting this interchange and evaluating the elementary inner integral we find

$$(25) \quad \int_{-2}^{2} A(1 + \epsilon + i\lambda y)\rho(y)e^{iuy}dy = \int_{0}^{\infty} (H(t) - 1)e^{-\epsilon t}\frac{\sin^2(u - \lambda t)}{(u - \lambda t)^2}dt$$

The left hand side of (25) is continuous in ϵ at $\epsilon = 0$, because A is holomorphic on the line $\Re z = 1$, and ρ is bounded. The limit as $\epsilon \searrow 0$ of the left hand side therefore exists. Also

$$\int_{0}^{\infty} \frac{\sin^2(u - \lambda t)}{(u - \lambda t)^2}dt < \infty$$

(because $\lambda > 0$ and u is real), so by the monotone convergence theorem we get from (25) that

$$\int_{0}^{\infty} H(t)\frac{\sin^2(u - \lambda t)}{(u - \lambda t)^2}dt < \infty$$

Going to the limit $\epsilon = 0$ in (25) we get

$$\int_{-2}^{2} A(1 + i\lambda y)\rho(y)e^{iuy}dy = \int_{0}^{\infty} (H(t) - 1)\frac{\sin^2(u - \lambda t)}{(u - \lambda t)^2}dt$$

The above holds for $\lambda > 0$ and any real u, so we may replace u by λu and change variable to get

$$\lambda \int_{-2}^{2} A(1 + i\lambda y)\rho(y)e^{i\lambda uy}dy = \int_{-\infty}^{\lambda u} \left(H\left(u - \frac{t}{\lambda}\right) - 1\right)\frac{\sin^2(t)}{t^2}dt$$

Writing the integral on the left hand side above as a sum of the integral over $[-2, 0]$ and that over $[0, 2]$ and integrating by parts we see that the integral on the left hand side is bounded by $C/|u|$ for a suitable constant $C = C(\lambda)$. In particular the left hand side converges to 0 as $u \to \infty$ (The reader may possibly recognize the statement as a special case of the Riemann-Lebesgue lemma). Thus for $\lambda > 0$,

$$(26) \quad \lim_{u \to \infty} \int_{-\infty}^{\lambda u} H\left(u - \frac{t}{\lambda}\right)\frac{\sin^2 t}{t^2}dt = \lim_{u \to \infty} \int_{-\infty}^{\lambda u} \frac{\sin^2 t}{t^2}dt = \int_{-\infty}^{\infty} \frac{\sin^2 t}{t^2}dt$$

It is not necessary to know the value of the last integral, so we just call it B (Actually $B = \pi$ by Exercise VI.16).

Observe that $e^t H(t) = \psi(e^t)$ is increasing in t, so that we for $b > 0$ have

$$(27) \quad e^t H(t) \le e^{t+b} H(t+b) \,, \text{ i.e. } H(t) \le H(t+b)e^b$$

Now, for $|t| < \sqrt{\lambda}$,

$$H(u) \le H\left(u + \frac{1}{\sqrt{\lambda}} - \frac{t}{\lambda}\right) \exp\left(\frac{1}{\sqrt{\lambda}} - \frac{t}{\lambda}\right) \le H\left(u + \frac{1}{\sqrt{\lambda}} - \frac{t}{\lambda}\right) \exp\left(\frac{2}{\sqrt{\lambda}}\right)$$

which implies that

$$\exp\left(-\frac{2}{\sqrt{\lambda}}\right) H(u) \int\limits_{-\sqrt{\lambda}}^{\sqrt{\lambda}} \frac{\sin^2 t}{t^2} dt \le \int\limits_{-\sqrt{\lambda}}^{\sqrt{\lambda}} H\left(u + \frac{1}{\sqrt{\lambda}} - \frac{t}{\lambda}\right) \frac{\sin^2 t}{t^2} dt$$

$$\le \int\limits_{-\infty}^{\lambda\left(u+\frac{1}{\sqrt{\lambda}}\right)} H\left(u + \frac{1}{\sqrt{\lambda}} - \frac{t}{\lambda}\right) \frac{\sin^2 t}{t^2} dt$$

Using (26) and letting $u \nearrow \infty$ above we get

$$\exp\left(-\frac{2}{\sqrt{\lambda}}\right) \limsup_{u\to\infty} H(u) \int\limits_{-\sqrt{\lambda}}^{\sqrt{\lambda}} \frac{\sin^2 t}{t^2} dt \le B$$

This relation holds for all $\lambda > 0$, so letting $\lambda \to \infty$ we deduce that

$$(28) \quad \limsup_{u\to\infty} H(u) \le 1$$

In particular H is bounded, say $H \le M$.

Let $\epsilon > 0$ be arbitrary, but fixed. For sufficiently large λ, say $\lambda > \Lambda(\epsilon)$ we have

$$\int\limits_{|t|>\sqrt{\lambda}} M\left(\frac{\sin t}{t}\right)^2 dt < \epsilon$$

Replacing u by $u - \frac{1}{\sqrt{\lambda}}$ in (26) we get for all $\lambda > \Lambda(\epsilon)$ that

$$B = \lim_{u\to\infty} \int\limits_{-\infty}^{\lambda\left(u-\frac{1}{\sqrt{\lambda}}\right)} H\left(u - \frac{1}{\sqrt{\lambda}} - \frac{t}{\lambda}\right) \frac{\sin^2 t}{t^2} dt$$

$$\le \epsilon + \liminf_{u\to\infty} \int\limits_{-\sqrt{\lambda}}^{\sqrt{\lambda}} H\left(u - \frac{1}{\sqrt{\lambda}} - \frac{t}{\lambda}\right) \frac{\sin^2 t}{t^2} dt$$

From (27) we get for $|t| < \sqrt{\lambda}$ that

$$(31) \quad H\left(u - \frac{1}{\sqrt{\lambda}} - \frac{t}{\lambda}\right) \leq H(u)e^{\frac{1}{\sqrt{\lambda}} + \frac{|t|}{\lambda}} \leq H(u)e^{\frac{2}{\sqrt{\lambda}}}$$

so we may continue the estimates as follows

$$B \leq \epsilon + \liminf_{u \to \infty} \int_{-\sqrt{\lambda}}^{\sqrt{\lambda}} H(u)e^{\frac{2}{\sqrt{\lambda}}} \frac{\sin^2 t}{t^2} dt$$

$$\leq \epsilon + e^{\frac{2}{\sqrt{\lambda}}} \liminf_{u \to \infty} H(u) \int_{-\sqrt{\lambda}}^{\sqrt{\lambda}} \frac{\sin^2 t}{t^2} dt \leq \epsilon + e^{\frac{2}{\sqrt{\lambda}}} \liminf_{u \to \infty} H(u)B$$

Letting $\lambda \to \infty$ we find $B \leq \epsilon + \liminf_{u \to \infty} H(u)B$, and so, $\epsilon > 0$ being arbitrary, $1 \leq \liminf_{u \to \infty} H(u)$. Comparing with (28) we have finally arrived at

Theorem 6 (The prime number theorem).
If $\pi(n)$ denotes the number of primes less than or equal to n, then

$$\frac{\pi(n)\log n}{n} \to 1 \quad as \quad n \to \infty$$

Section 5 Exercises

1. Show that

$$\sum_p \frac{1}{p} = \infty$$

where the summation ranges over all primes p. Hint: If not, then the infinite product

$$\prod_p \left(1 - \frac{1}{p}\right)$$

converges, and you may take $z = 1$ in the procedure of the proof of Theorem 3.

2. Let $\mu : \mathbf{N} \to \mathbf{R}$ be the *Möbius function*, i.e.

$\mu(1) = 1$
$\mu(p_1^{\alpha_1} \cdots p_k^{\alpha_k}) = (-1)^k$ if $p_1, \cdots, p_k$ are distinct primes and $\alpha_1 = \cdots = \alpha_k = 1$
$\mu(n) = 0$ otherwise

Show that the series (a so-called *Dirichlet series*)

$$\sum_{n=1}^{\infty} \frac{\mu(n)}{n^z}$$

converges absolutely when $\Re z > 1$ and that

$$\zeta(z) \sum_{n=1}^{\infty} \frac{\mu(n)}{n^z} = 1$$

when $\Re z > 1$.

Hint: Note first that

$$\sum_{d|n} \mu(d) = \begin{cases} 1 & \text{if } n = 1 \\ 0 & \text{if } n > 1 \end{cases}$$

3. Let $\phi : \mathbf{N} \to \mathbf{C}$ have the properties

(i) $\phi(nm) = \phi(n)\phi(m)$ whenever $(n, m) = 1$

(ii) $\sum_{n=1}^{\infty} |\phi(n)| < \infty$

Show that

$$\sum_{n=1}^{\infty} \phi(n) = \prod_{p} \left\{ 1 + \phi(p) + \phi(p^2) + \cdots \right\}$$

where p in the product ranges over all primes.

Apply this result to $\phi(n) := \mu(n)/n^z$, where $z \in \mathbf{C}$ is fixed with $\Re z > 0$ and μ is the Möbius function from the previous exercise, to get

$$\sum_{n=1}^{\infty} \frac{\mu(n)}{n^z} = \prod_{p} \left(1 - p^{-z} \right)$$

As another application take $\phi = 1$.

4. The result

$$2 \int_{0}^{\infty} \frac{u^z}{1 + u^2} du = \frac{\pi}{\cos\left(\frac{\pi z}{2}\right)} \quad \text{for } -1 < \Re z < 1$$

from the proof of Theorem 2 can also be verified by calculus of residues. Do it!

Chapter 12 Harmonic Functions

Harmonic functions are solutions to the Laplace equation $\Delta u = 0$, where

$$\Delta = \frac{\partial^2}{\partial x^2} + \frac{\partial^2}{\partial y^2}$$

The Laplace operator Δ crops up in so many practical and theoretical situations, that it is hard to overestimate its importance. In the context of holomorphic functions every harmonic function is the real part of a holomorphic function in a simply connected domain.

Subharmonic functions, which are the topic for the next chapter, are very useful in the study of holomorphic functions. It may be said that this largely is due to the fact that when f is holomorphic then $\log |f(z)|$ is subharmonic. We will approach this via Jensen's formula. Harmonic and subharmonic functions are very closely related in much the same way as linear and convex functions.

Section 1 Holomorphic and harmonic functions

Definition 1.

Let Ω be an open subset of $\mathbf{R}^n$. A real valued function $u \in C^2(\Omega)$ is said to be harmonic *if $\Delta u = 0$, where*

$$\Delta = \frac{\partial^2}{\partial x_1^2} + \frac{\partial^2}{\partial x_2^2} + \cdots + \frac{\partial^2}{\partial x_n^2}$$

is the so-called Laplace *operator.*

The Laplace equation $\Delta u = 0$ occurs not just in mathematics, but also in many different connections in physics and mechanics, in particular in descriptions of stationary situations and equilibrium states.

Theorem 2.

Let Ω be an open subset of $\mathbf{R}^n$.

(α) If $f \in Hol(\Omega)$ then $\Re f$ and $\Im f$ are harmonic in Ω.

(β) Assume that Ω is simply connected. If u is harmonic in Ω, then there exists an $f \in Hol(\Omega)$ such that $u = \Re f$.

The theorem yields a lot of interesting harmonic functions, e.g.

$$u(x,y) = Log(x^2 + y^2) \quad \text{defined for} \quad (x,y) \in \mathbf{R}^2 \backslash \{0,0\}$$

183

Proof of Theorem 2:

(α) The Cauchy-Riemann equations tell us that $\partial f/\partial \bar{z} = 0$, so the result is a consequence of the formula

$$\Delta f = 4 \frac{\partial}{\partial z} \left(\frac{\partial f}{\partial \bar{z}} \right)$$

(β) The function

$$h := \frac{\partial u}{\partial x} - i \frac{\partial u}{\partial y} : \Omega \to \mathbf{C}$$

is C^1 in Ω and satisfies the Cauchy-Riemann equations, so it is holomorphic in Ω. Since Ω is simply connected h has a primitive f (Theorem V.8), and (again by the Cauchy-Riemann equations)

$$h = f' = \frac{\partial(\Re f)}{\partial x} - i \frac{\partial(\Re f)}{\partial y}$$

Comparing with the definition of h we get

$$\frac{\partial u}{\partial x} = \frac{\partial(\Re f)}{\partial x} \quad \text{and} \quad \frac{\partial u}{\partial y} = \frac{\partial(\Re f)}{\partial y}$$

so $u = \Re f +$ some constant. When we absorb the constant in f the proof is complete.□

Corollary 3. *A harmonic function is C^∞.*

The function $u(x, y) = Log(x^2 + y^2)$ is not the real part of an analytic function in all of $\mathbf{C}\backslash\{0\}$: The analytic function in question would be $2 \log z$, which, as we know, cannot be defined in all of $\mathbf{C}\backslash\{0\}$. However that is the only regrettable thing that may happen:

Theorem 4 (The Logarithmic Conjugation Theorem).
Let K be a connected compact subset of $B(0, R)$, where $R \in \,]0, \infty]$. Let $z_0 \in K$. If u is harmonic on $\Omega := B(0, R)\backslash K$ then there exist an analytic function f on Ω and a real constant c such that

$$u(z) = \Re f(z) + cLog|z - z_0| \text{ for all } z \in \Omega$$

The applications of the Logarithmic Conjugation Theorem are typically to ring shaped domains, i.e. the K of the theorem is a closed ball, possibly degenerated to a point.

Proof: As in the proof of Theorem 2(α) we note that

$$h := \frac{\partial u}{\partial x} - i \frac{\partial u}{\partial y} : \Omega \to \mathbf{C}$$

is holomorphic in Ω, because h satisfies the Cauchy-Riemann equations.

Choose $r \in]0, R[$ such that $K \subseteq B(0, r)$, and let C denote the circle $|z - z_0| = r$. Let us from now on take $z_0 = 0$ for simplicity.

Let γ be any closed path in Ω. If we put $m := Ind_\gamma(z_0)$, then $Ind_\gamma(z) = m$ for all $z \in K$, K being connected. Now,

$$Ind_\gamma(z) = m \, Ind_C(z) \text{ for all } z \in \mathbf{C} \backslash \Omega$$

so by the global Cauchy Theorem (Corollary V.4) we get for any complex number c that

$$\int_\gamma \left(h(z) - \frac{c}{z} \right) dz = m \int_C \left(h(z) - \frac{c}{z} \right) dz = m \left(\int_C h(z) dz - 2\pi i c \right)$$

Choosing

$$c := \frac{1}{2\pi i} \int_C h$$

we get

$$\int_\gamma \left(h(z) - \frac{c}{z} \right) dz = 0 \text{ for all closed paths } \gamma \text{ in } \mathbf{C} \backslash \{0\}$$

It follows that $z \to h(z) - c/z$ has a primitive in each connected component of Ω, hence in all of Ω. Letting f denote such a primitive we have $f'(z) = h(z) - c/z$. We shall soon need that c is real, so we prove that now:

$$c = \frac{1}{2\pi i} \int_C h(z) dz = \frac{1}{2\pi} \int_0^{2\pi} h\left(e^{i\phi}\right) e^{i\phi} d\phi = \frac{1}{2\pi} \int_0^{2\pi} \left\{ \frac{\partial u}{\partial x}\left(e^{i\phi}\right) - i\frac{\partial u}{\partial y}\left(e^{i\phi}\right) \right\} d\phi$$

$$= \frac{1}{2\pi} \int_0^{2\pi} \left\{ \frac{\partial u}{\partial x} \cos\phi + \frac{\partial u}{\partial y} \sin\phi + i\frac{\partial u}{\partial x} \sin\phi - i\frac{\partial u}{\partial y} \cos\phi \right\} d\phi$$

so

$$\Im c = \frac{1}{2\pi} \int_0^{2\pi} \left\{ \frac{\partial u}{\partial x} \sin\phi - \frac{\partial u}{\partial y} \cos\phi \right\} d\phi$$

$$= -\frac{1}{2\pi} \int_0^{2\pi} \frac{d}{d\phi} \{u(\cos\phi, \sin\phi)\} d\phi = 0$$

i.e. c is real.

Now,

$$\frac{\partial}{\partial x} \{\Re f + c Log|z|\} = \Re \frac{\partial}{\partial x} \{f + c Log|z|\} = \Re \left\{ f' + \frac{c}{z} \right\} = \Re h = \frac{\partial u}{\partial x}$$

185

and similarly

$$\frac{\partial}{\partial y}\{\Re f + cLog|z|\} = \frac{\partial u}{\partial y}$$

so that

$$\Re f + cLog|z| = u + g$$

where g is constant on each connected component of Ω. Absorbing the constant in f the proof is finished. $\qquad\qquad\square$

In the above version of the Logarithmic Conjugation Theorem the complement of Ω has only one bounded connected component, viz. K. In general each bounded connected component contributes with a logarithmic term.

Let us use the Logarithmic Conjugation Theorem to prove the following classical result. A simple proof, that uses the maximum principle, but not the relationship between harmonic and analytic functions, can be found in [Pe;Satz III.30.6].

Theorem 5.

If a harmonic function is bounded in a neighborhood of an isolated singularity then that singularity is a removable singularity.

Proof : We may assume that the harmonic function u is bounded in a punctured disc D' around 0. By the Logarithmic Conjugation Theorem u has the form

$$u(z) = \Re f(z) + cLog|z| \text{ where } f \in Hol(D') \text{ and } c \in \mathbf{R}$$

We will first show that $c = 0$. If not we may divide through by it and so assume that $c = 1$. Now, $Log|z| \to -\infty$ as $z \to 0$, so to balance it $\Re f(z) \to \infty$ as $z \to \infty$. In particular we see that $|f(z)| \to \infty$ as $z \to 0$, i.e. f has a pole at $z = 0$, and so f may be written in the form

$$f(z) = \frac{h(z)}{z^N}$$

where $N \geq 1$ and $h(0)$ is not 0.

Choosing z_n as an Nth root of $-h(0)/n$ we find

$$f(z_n) = -\frac{h(z_n)}{h(0)}$$

so

$$\Re f(z_n) = -\frac{\Re h(z_n)}{h(0)} \to -\infty$$

But that contradicts $\Re f(z) \to \infty$ as $z \to 0$, so $c = 0$, and thus $u = \Re f$.

If we assume that f has a pole at 0 we may derive a contradiction in the same way as above. If we assume that 0 is an essential singularity then we get a contradiction by Picard's big theorem or just by Casorati-Weierstrass (Remember $\Re f = u$ is bounded). So left is the desired case of 0 being a removable singularity for f. $\square$

For information on the Logarithmic Conjugation Theorem in finitely connected regions and its applications we recommend the paper [Ax] from where the above treatment is taken.

Section 2 Poisson's formula

The important Poisson formula expresses harmonic functions in terms of their boundary values:

Theorem 6. (Poisson's Formula).
If u is continuous in $B[0, R]$ and harmonic in $B(0, R)$ then

$$u(z) = \frac{1}{2\pi} \int_0^{2\pi} u\left(Re^{i\phi}\right) \frac{R^2 - |z|^2}{\left|Re^{i\phi} - z\right|^2} d\phi \ \ for \ |z| < R$$

In particular u has the mean value property, i.e.

$$u(0) = \frac{1}{2\pi} \int_0^{2\pi} u\left(Re^{i\phi}\right) d\phi$$

Proof: The proof is from [Aa]. We will assume that u is harmonic in an open ball containing $B[0, R]$ and leave the derivation of the general case to the reader.

Let $|z| < R$. The Cauchy integral formula for the function

$$z \rightarrow \frac{f(z)}{R^2 - z\overline{a}}$$

in the disc $B[0, R]$ is

$$\frac{f(z)}{R^2 - z\overline{a}} = \frac{1}{2\pi i} \int_{|w|=R} \frac{f(w)}{R^2 - w\overline{a}} \frac{1}{w - z} dw$$

$$= \frac{1}{2\pi} \int_0^{2\pi} \frac{f\left(Re^{i\phi}\right)}{R^2 - Re^{i\phi}\overline{a}} \frac{Re^{i\phi}}{Re^{i\phi} - z} d\phi = \frac{1}{2\pi} \int_0^{2\pi} \frac{f\left(Re^{i\phi}\right)}{Re^{-i\phi} - \overline{a}} \frac{1}{Re^{i\phi} - z} d\phi$$

Taking $z = a$ here we get Poisson's formula. $\square$

187

If we as center of the circle take z instead of 0 then the resulting formula is

$$u(z) = \frac{1}{2\pi} \int_0^{2\pi} u\left(z + Re^{i\phi}\right) d\phi$$

i.e. the value of u at the center of a ball is the mean value of the values of u at the boundary. We say that u has the *mean value property*.

The function

$$P_R(\phi, z) := \frac{1}{2\pi} \frac{R^2 - |z|^2}{|Re^{i\phi} - z|^2} \ \ for \ \ \phi \in \mathbf{R} \ \ and \ \ |z| < R$$

is called the *Poisson kernel for B(0,R)* and is of great value in the study of harmonic functions. We list a couple of its properties that will be useful in the sequel:

Properties of the Poisson kernel:
(i)
$$P_R > 0$$

(ii)
$$\int_0^{2\pi} P_R(\phi, z) d\phi = 1 \ \ when \ \ |z| < R$$

(iii)
$$P_R(\phi, z) = \Re\left\{ \frac{1}{2\pi} \frac{Re^{i\phi} + z}{Re^{i\phi} - z} \right\} = \frac{1}{2\pi} \Re\left\{ 1 + 2 \sum_{n=1}^{\infty} R^{-n} e^{-in\phi} z^n \right\}$$

In particular $P_R(\phi, z)$ is harmonic in $B(0, R)$ for fixed ϕ.
(iv)
$$\frac{1}{2\pi} \frac{R - |z|}{R + |z|} \leq P_R(\phi, z) \leq \frac{1}{2\pi} \frac{R + |z|}{R - |z|}$$

(v) If $0 < \delta < \pi/2$ then

$$P_R(\phi, z) \leq \frac{1 - R^2}{\sin^2(\delta)} \ \ for \ \delta \leq \phi \leq 2\pi - \delta \ and \ |z| < R$$

Proof: The proofs of (i), (iii), (iv) and (v) are elementary estimates and computations left to the reader. You get (ii) by putting $u = 1$ in Poisson's formula. $\square$

The following theorem is a consequence of the mean value property. Maximum principles are among the most useful tools employed in differential equations. See [PW].

Theorem 7 (The strong maximum principle).

Let u be harmonic in an open connected subset Ω of $\mathbf{R}^2$, and assume that u takes its supremum, i.e. there exists a $z_0 \in \Omega$ such that $u(z_0) = \sup\{u(z)|z \in \Omega\}$.

Then u is constant.

Proof: It suffices to show that the set $\Omega_0 := \{z \in \Omega \,|\, u(z) = u(z_0)\}$ is both closed and open in Ω. It is closed because u is continuous. Let $z \in \Omega_0$; for the simplicity of writing , assume that $z = 0$. If $B[0, R] \subseteq \Omega$, then we get from the mean value property for any $0 < r \le R$ that

$$0 = u(0) - \frac{1}{2\pi} \int\limits_0^{2\pi} u\left(re^{i\phi}\right) d\phi = \frac{1}{2\pi} \int\limits_0^{2\pi} \left\{ u(0) - u\left(re^{i\phi}\right) \right\} d\phi$$

and - since $u(0) \ge u\left(re^{i\phi}\right)$ - we get $u(0) = u(re^{i\phi})$ for all ϕ. So $u(0) = u(w)$ for all $w \in B(0, R)$. This shows that the set Ω_0 is open. $\qquad\square$

We shall next formulate an inverse to Poisson's formula:

Theorem 8.

Let f be continuous and real valued on the circle $|z| = R$. Then the function u, defined by

$$u(z) := \begin{cases} \int\limits_0^{2\pi} f(Re^{i\phi}) P_R(\phi, z) d\phi & \text{for } |z| < R \\ f(z) & \text{for } |z| = R \end{cases}$$

is continuous in the closed ball $B[0, R]$ and harmonic in the open ball $B(0, R)$.

Proof: The procedure of proof is borrowed from [Mi].

We treat only the case $R = 1$ and leave the derivation of the general case to the reader. For $|z| < 1$ we have

$$u(z) = \int\limits_0^{2\pi} f\left(e^{i\phi}\right) P_1(\phi, z) d\phi = \Re \int\limits_0^{2\pi} f\left(e^{i\phi}\right) \frac{e^{i\phi} + z}{e^{i\phi} - z} \, d\phi$$

The last integral can be expanded in a power series in z, so u is the real part of an analytic function and is hence harmonic.

Left is continuity at the boundary $|z| = 1$.

To motivate the procedure we recall the Möbius transformation $A(z; w)$ from Example VIII.14. If our theorem is correct then the mean value property for the

189

harmonic function $u \circ A(-z, \cdot)$ implies that

$$\int_0^{2\pi} f(e^{it}) P_1(t,z) dt = u(z) = [u \circ A(-z, \cdot)](0)$$

$$= \frac{1}{2\pi} \int_0^{2\pi} u\left(A\left(-z, e^{i\theta}\right)\right) d\theta = \frac{1}{2\pi} \int_0^{2\pi} f\left(\frac{e^{i\theta} + z}{1 + \bar{z}e^{i\theta}}\right) d\theta$$

The crucial point of the proof is to establish the identity

$$(*) \quad \int_0^{2\pi} f(e^{it}) P_1(t,z) dt = \frac{1}{2\pi} \int_0^{2\pi} f\left(\frac{e^{i\theta} + z}{1 + \bar{z}e^{i\theta}}\right) d\theta \text{ for } |z| < 1$$

To do so we note that the map

$$w \to A\left(z; e^{iw}\right) \frac{1 - \bar{z}}{1 - z}$$

is analytic and nowhere vanishing on a sufficiently small neighborhood of $[0, 2\pi]$, so it has a continuous and hence analytic logarithm there, say $i\Theta$. We may choose Θ so that $\Theta(0) = 0$. Letting $\theta := \Theta|[0, 2\pi]$ we have the identity

$$e^{i\theta(t)} = \frac{e^{it} - z}{1 - \bar{z}e^{it}} \frac{1 - \bar{z}}{1 - z}$$

A differentiation of this identity with respect to t shows that

$$\theta'(t) = \frac{1 - |z|^2}{|e^{it} - z|^2} = 2\pi P_1(t, z)$$

From property (ii) of the Poisson kernel we infer that θ is an increasing bijection of $[0, 2\pi]$ onto $[0, 2\pi]$. And when we introduce t as new variable on the right hand side of (*) we get (*) by the change of variables formula.

Since the restriction of u to the boundary is continuous (being equal to f there), it suffices to finish the proof of the theorem to show that $u(z_n) \to f(e^{i\theta_0})$ for any sequence $\{z_n\}$ from the open ball $B(0, 1)$ converging to the arbitrary point $e^{i\theta_0}$ of the boundary. But f is continuous, hence bounded, so this is an immediate consequence of (*) and Lebesgue's dominated convergence theorem. □

At this point we must mention the

Dirichlet problem.

Given an open subset Ω of $\mathbf{R}^n$ and a continuous function f on the boundary $\partial\Omega$. Find a function u such that u is continuous on the closure $\overline{\Omega}$ of Ω, $\triangle u = 0$ in Ω and $u|_{\partial\Omega} = f$.

The Dirichlet problem in $\mathbf{R}^3$ may be interpreted as finding the electrostatic potential u inside a hollow cavity Ω, given the potential at the boundary $\partial\Omega$.

Proposition 9.

Let Ω be a bounded open subset of $\mathbf{R}^2$. Then the Dirichlet problem has at most one solution, given the boundary values.

Proof: We shall show that $u \in C(\Omega)$, u is harmonic in Ω and $u|_{\partial\Omega} = 0$ implies that $u = 0$.

Since $\overline{\Omega}$ is compact, u assumes its supremum somewhere in $\overline{\Omega}$, say at $z_0 \in \Omega$. If $z_0 \in \partial\Omega$ then $u \le 0$ $(= u(z_0))$ everywhere. If $z_0 \in \Omega$ then u is by the maximum principle constant throughout the connected component of z_0, so the value $u(z_0)$ is also taken at some boundary point. Thus $u \le 0$ once again.

Replacing u by $-u$ we get $u = 0$. □

Theorem 8 shows that the Dirichlet problem for a disc has a solution. Even better, it provides us with an explicit formula for the (unique) solution.

Theorem 10 (Harnack's monotone convergence theorem).

Let Ω be an open connected subset of $\mathbf{R}^2$. Let $u_1 \le u_2 \le \cdots \le u_n \le \cdots$ be an increasing sequence of functions that are harmonic in Ω. Assume that there exists a point $z_0 \in \Omega$ such that $\sup\{u_n(z_0)\,|\,n = 1, 2, \cdots\} < \infty$.

Then $\{u_n\}$ converges locally uniformly in Ω to a harmonic function.

Proof:

Let $B[z_0, R] \subseteq \Omega$. We will for $n, p \in \mathbf{N}$ estimate the positive function $v_{n,p} := u_{n+p} - u_n$ in $B[z_0, R]$. Choose $\epsilon > 0$ such that $B[z_0, R + \epsilon] \subseteq \Omega$, and let us for the sake of simplicity assume that $z_0 = 0$. From the Poisson formula we get for any $z \in B(0, R + \epsilon)$ that

$$v_{n,p}(z) = \int_0^{2\pi} v_{n,p}\Big((R + \epsilon)e^{i\phi}\Big) P_{R+\epsilon}(\phi, z)d\phi$$

and so, from Property (iv) of the Poisson kernel, the estimates

$$\frac{1}{2\pi} \int_0^{2\pi} v_{n,p}\Big([R + \epsilon]e^{i\phi}\Big) \frac{R + \epsilon - |z|}{R + \epsilon + |z|} d\phi \le v_{n,p}(z)$$

$$\le \frac{1}{2\pi} \int_0^{2\pi} v_{n,p}\Big([R + \epsilon]e^{i\phi}\Big) \frac{R + \epsilon + |z|}{R + \epsilon - |z|} d\phi$$

191

which by the mean value property reduce to

$$\frac{R + \epsilon - |z|}{R + \epsilon + |z|} v_{n,p}(0) \leq v_{n,p}(z) \leq \frac{R + \epsilon + |z|}{R + \epsilon - |z|} v_{n,p}(0)$$

These inequalities show that $\{u_n\}$ converges uniformly in $B[z_0, R]$.

The argument actually shows that if $\{u_n\}$ converges in a point z, then it converges uniformly in any closed ball $B[z, R]$ in Ω. Thus the set $\{z \in \Omega | \{u_n(z)\}$ converges$\}$ is both open and closed in Ω, and so by connectedness it equals Ω or the empty set. But z_0 belongs to the set.

We have now shown that $\{u_n\}$ converges uniformly on any closed ball in Ω. Since any compact set in Ω can be covered by finitely many such balls the convergence is uniform on compact subsets of Ω.

Since u_n is continuous, so is the limit function u.

Let $z_1 \in \Omega$ and $B[z_1, R] \subseteq \Omega$. Letting $n \to \infty$ in the formula

$$u_n(z) = \int_0^{2\pi} u_n \Big(z_1 + R e^{i\phi} \Big) P_R(\phi, z - z_1) d\phi$$

we get for z in the ball $B(z_1, R)$ that

$$u(z) = \int_0^{2\pi} u \Big(z_1 + R e^{i\phi} \Big) P_R(\phi, z - z_1) d\phi$$

It now follows from Theorem 8 that u is harmonic in $B(z_1, R)$. But z_1 was arbitrary in Ω. $\qquad\qquad\qquad\square$

Section 3 Jensen's formula

If f is holomorphic and without zeros in an open set containing the ball $B[0, R]$ then $\log|f|$ is harmonic inside the ball, so by the mean value property

$$\log|f(0)| = \frac{1}{2\pi} \int_0^{2\pi} \log\Big| f\Big(R e^{i\phi} \Big) \Big| d\phi$$

In the case where f has zeros *Jensen's inequality*

$$\log|f(0)| \leq \frac{1}{2\pi} \int_0^{2\pi} \log\Big| f\Big(R e^{i\phi} \Big) \Big| d\phi$$

holds as a consequence of Jensen's formula (Theorem 11 below). Jensen's inequality also follows from the theory of subharmonic functions (next chapter).

Jensen's formula tells what the difference is between the right hand side and the left hand side of Jensen's inequality. It is useful tool for providing information on the zeros of analytic functions on a disc, when restrictions on their sizes are imposed.

Theorem 11 (Jensen's formula).

Let f be meromorphic in an open set containing $B[0, R]$. Let $z_1, z_2, \cdots, z_n$ and $p_1, p_2, \cdots, p_m$ be the zeros and poles respectively of f in $B(0, R)$ (counted with multiplicity). Assume none of them is zero.

Then $\phi \to \log\left|f(Re^{i\phi})\right|$ is integrable over $(0, 2\pi)$, and

$$\log|f(0)| + \log\left|R^{n-m}\frac{p_1 \cdots p_m}{z_1 \cdots z_n}\right| = \frac{1}{2\pi}\int_0^{2\pi} \log\left|f(Re^{i\phi})\right| d\phi$$

Proof: Let $z_1, z_2, \cdots, z_n, \cdots, z_N$ and $p_1, p_2, \cdots, p_m, \cdots, p_M$ be the zeros and poles respectively of f in the closed ball $B[0, R]$, counted with multiplicity. The function

$$F(z) := f(z) \prod_{j=1}^{m} \frac{R(z - p_j)}{R^2 - \overline{p_j}} \prod_{j=m+1}^{M} \frac{p_j - z}{p_j} \left\{ \prod_{k=1}^{n} \frac{R(z - z_k)}{R^2 - \overline{z_k}z} \prod_{k=n+1}^{N} \frac{z_k - z}{z_k} \right\}^{-1}$$

is holomorphic and zero free in a neighborhood of $B[0, R]$, so

$$\log|F(0)| = \frac{1}{2\pi}\int_0^{2\pi} \log\left|F(Re^{i\phi})\right| d\phi$$

Putting $z = 0$, we find from the expression for F that

$$|F(0)| = |f(0)| \prod_{j=1}^{m} \frac{|p_j|}{R} \prod_{k=1}^{n} \frac{R}{|z_k|}$$

so $\log|F(0)|$ is the left hand side of the desired formula.

Writing $p_j = Re^{i\theta_j}$ for $j = m+1, \cdots, M$ and $z_k = Re^{i\phi_k}$ for $k = n+1, \cdots, N$ we find that

$$\left|F\left(Re^{i\theta}\right)\right| = \left|f\left(Re^{i\theta}\right)\right| \prod_{j=m+1}^{M} \left|1 - e^{i(\theta-\theta_j)}\right| \prod_{k=n+1}^{N} \left|1 - e^{i(\phi-\phi_k)}\right|$$

so taking logarithms we get

$$\log\left|F\left(Re^{i\theta}\right)\right| = \log\left|f\left(Re^{i\theta}\right)\right|$$
$$+ \sum_{j=m+1}^{M} \log\left|1 - e^{i(\theta-\theta_j)}\right| - \sum_{k=n+1}^{N} \log\left|1 - e^{i(\phi-\phi_k)}\right|$$

To finish the proof it now suffices to show that $\phi \to \log\left|1 - e^{i\phi}\right|$ is integrable over $]-\pi, \pi[$ and that

$$\int_{-\pi}^{\pi} \log\left|1 - e^{i\phi}\right| d\phi = 0$$

(Now, this is a book on complex function theory, so we shall here evaluate the integral by help of the Cauchy theorem. But the integral can actually be computed by elementary means [AGR],[Yo]).

Since $x \to \log|x|$ is locally integrable, and the function $(1 - e^{i\phi})/\phi$ is continuous and never 0 on $[-\pi, \pi]$, the function

$$\phi \to \log\left|1 - e^{i\phi}\right| = \log\left|\frac{1 - e^{i\phi}}{\phi}\right| + \log|\phi|$$

is integrable over $]-\pi, \pi[$.

To show that the value of the integral is 0 we note that

$$z \to \frac{Log\ z}{z - 1}$$

is analytic in the right half plane, so that its integral along any circle $|z - 1| = r$, where $r \in]0, 1[$, is zero by the Cauchy theorem, i.e.

$$0 = \frac{1}{i} \int_{|z-1|=r} \frac{Log\ z}{z - 1} dz = \int_{-\pi}^{\pi} Log\left(1 - re^{i\phi}\right) d\phi$$

Taking real parts we get

$$\int_{-\pi}^{\pi} \log\left|1 - re^{i\phi}\right| d\phi = 0$$

We will apply the dominated convergence theorem to this for a sequence $r_n \uparrow 1$. The desired result is then an immediate consequence, once we dominate the integrands by an integrable function.

By differentiation we see that the function

$$\phi \to \left|\frac{1 - re^{i\phi}}{1 - e^{i\phi}}\right|$$

for fixed r is decreasing for $\phi > 0$, so

$$\frac{2}{\left|1 - e^{i\phi}\right|} > \left|\frac{1 - re^{i\phi}}{1 - e^{i\phi}}\right| \geq \frac{1 + r}{2} > \frac{1}{2}$$

and so

$$\log 2 > \log \left|1 - re^{i\phi}\right| \geq \log \left|1 - e^{i\phi}\right| - \log 2$$

$\square$

Section 4 Exercises

1. Let u be harmonic in Ω. Show that $u(\Omega)$ is open, unless u is constant. Hint: Connected subsets of the real line are intervals.

2. Derive the maximum principle for holomorphic functions from the strong maximum principle for harmonic functions.

3. Let f be a continuous, bounded and real valued function on the real line. Define for $x \in \mathbf{R}$ and $y > 0$ a function u by

$$u(x + iy) := \begin{cases} \frac{1}{\pi} \int_{-\infty}^{\infty} \frac{yf(t)}{(x-t)^2 + y^2} dt & \text{for } y > 0 \\ f(x) & \text{for } y = 0 \end{cases}$$

(α) Show that u is bounded and harmonic in the upper half plane.

(β) Show that u is continuous in the closed upper half plane.

4. Let u be harmonic in Ω and identically 0 in an open, non-empty subset of Ω. Show that $u = 0$ in all of Ω. (This property of harmonic functions is called the *weak unique continuation property*). Hint: The unique continuation for holomorphic functions.

5. Define the map $u : \{x + iy \in \mathbf{C} \,|\, y > 0\} \to]0, \pi[$ by

$u(z) :=$ the angle under which the interval $[0,1]$ is seen from z.

Show that u is harmonic. Hint: Consider the function $Log((z - 1)/z)$.

6. Show that the function

$$u(x, y) := \frac{y}{x^2 + y^2}$$

is harmonic on $\mathbf{R}^2$ except at the origin.

7. Is $|z|^n$ harmonic for some n?

8. Show that the polynomial $u(x, y) := x^3 - 3xy^2 - x$ is harmonic on $\mathbf{R}^2$. Find an $f \in Hol(\mathbf{C})$ such that $u = \Re f$.

Same problems with $u(x, y) := \sin x \cosh y$

9. Show the following formula for $r \in [0, 1[$ and $\theta \in \mathbf{R}$:

$$r \cos \theta = \frac{1}{2\pi} \int_{-\pi}^{\pi} \frac{(1 - r^2) \cos \phi}{1 - 2r \cos(\phi - \theta) + r^2} d\phi$$

Hint: Solve the Dirichlet problem when the boundary function is $f(x + iy) = x$.

10. (Liouville's theorem for harmonic functions) Let u be harmonic on $\mathbf{R}^2$ and assume that u is bounded from below. Show that u is a constant. Hint: Property (iv) of the Poisson kernel. See also [Bo] and [Ch].

11. Show that the Dirichlet problem on $\{z \in \mathbf{C} \,|\, 0 < |z| < 1\}$ with the boundary function $f(z) = 0$ for $|z| = 1$, $f(0) = 1$ has no solution.

12. Let u be harmonic in the annulus $\{z \in \mathbf{C} \,|\, R_1 < |z| < R_2\}$. Show that

$$\frac{1}{2\pi} \int_0^{2\pi} u\left(re^{i\phi}\right) d\phi$$

is a linear function of $\log r$, i.e. there are real constants c and d such that for all $r \in]R_1, R_2[$:

$$\frac{1}{2\pi} \int_0^{2\pi} u\left(re^{i\phi}\right) d\phi = c \log r + d$$

Hint: The Logarithmic Conjugation Theorem.

13. Let $R \in]0, \infty[$. *Inversion* in the circle $|z| = R$ is the map

$$p_R : \mathbf{R}^2 \backslash \{0\} \to \mathbf{R}^2 \backslash \{0\}$$

defined by

$$p_R(z) := \frac{R^2}{|z|^2} z \quad \text{for } z \in \mathbf{R}^2 \backslash \{0\}$$

Note that $p_R(z) = z$ when $|z| = R$, that points inside the circle are mapped outside and vice versa, and that

$$p_R \circ p_R = identity \text{ on all of } \mathbf{R}^2 \backslash \{0\}$$

(α) Let u be harmonic in $\mathbf{R}^2 \backslash \{0\}$. Show that so is $u \circ p_R$.

(β) Define and prove similar statements for the circle $|z - z_0| = R$ centered at $z_0 \in \mathbf{R}^2$.

14. Let u be harmonic in all of $\mathbf{R}^2$, positive on the open upper half space and 0 on the real axis.

(α) Let z_0 be a point in the open lower half plane. Let p denote inversion in the circle $|z - z_0| = R$ (See the previous exercise). Let z be a point in the open upper half plane such that $|z - z_0| < R$. Show that $u(z) \leq u(p(z))$.

Hint: Apply the weak maximum principle to $u \circ p - u$ in the domain D indicated on the figure

196

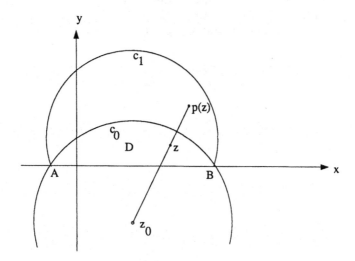

The curve c_1 is the image by p of the interval [A,B].

(β) Let $0 < y_1 < y_2$. Show that

$$u(x_1, y_1) \leq u(x_2, y_2) \text{ for all } x_1, x_2 \in \mathbf{R}$$

(γ) Show that $u(x, y)$ is independent of x for $y > 0$.

(δ) Show that $u(x, y) = cy$ where c is a constant.

(ϵ) Let f be an entire function that maps the upper half plane into itself and is real on the real line. Show that f is an affine function, i.e. has the form $f(z) = az + b$ where a and b are constants.

The exercise is taken from [Ti]. See also [Bo].

15. Prove the following theorem, due to Kellogg. It says that in special situations to get harmonicity we do not need the mean value property for all sufficiently small circles around the points of the domain, but only for one circle around each point.

Theorem (Kellogg's Theorem).

*Let $u \in C(B[0,1])$ and assume that there for each $z \in B(0,1)$ exists an $r = r(z) \in$
$]0, 1 - |z|[$ such that*

$$u(z) = \frac{1}{2\pi} \int\limits_{-\pi}^{\pi} u\left(z + re^{i\phi}\right) d\phi$$

Then u is harmonic on $B(0,1)$.

Hints to the proof: Show that we may assume $u = 0$ on the unit circle by subtracting a suitable harmonic function. Assume $u(z) > 0$ for some $z \in B(0,1)$;

obtain a contradiction by applying the assumption of the theorem to a point in $\{z \in B[0,1] \mid u(z) = \|u\|_\infty\}$ closest to the boundary.

16. As a curiosity derive the fundamental theorem of algebra from Jensen's formula.

Chapter 13 Subharmonic Functions

Section 1 Technical results on upper semicontinuous functions

§1. Technical results on upper semicontinuous functions.

Definition 1.

Let X be a topological space. A function $u : X \to [-\infty, \infty[$ is said to be upper semi-continuous *(usc) if the following holds :*

To every $x_0 \in X$ and $M > u(x_0)$ there exists a neighborhood U of x_0 such that $M > u(x)$ for all $x \in U$.

A continuous real valued function is clearly usc. Note for later use that an usc function is Borel measurable - indeed, the set $\{z \in \Omega | u(z) < \alpha\}$ is open for any $\alpha \in \mathbf{R}$.

Lemma 2.

If u_1 and u_2 are usc then so are $u_1 + u_2$ and $\max\{u_1, u_2\}$.

If u is usc and $\lambda \in [0, \infty[$ then λu is usc.

If u_α is usc for each α in a non-empty index set A then $\inf\limits_{\alpha \in A} u_\alpha$ is also usc.

Theorem 3.

An upper semi-continuous function on a compact topological space attains its supremum. In particular it is bounded from above.

Proof : As the standard proof for continuous functions. □

Theorem 4.

Let $u : \Omega \to [-\infty, \infty[$ where Ω is an open subset of the complex plane.

Then u is usc if and only if there exists a decreasing sequence $u_1 \geq u_2 \geq \cdots$ of continuous real valued functions on Ω such that $u_n(z) \to u(z)$ as $n \to \infty$ for each $z \in \Omega$.

Proof : The theorem is trivial in case $u(z) = -\infty$ for all z, so we will from now on assume that there exists a $z_0 \in \Omega$ such that $u(z_0) \in \mathbf{R}$.

The if direction of the theorem is part of the lemma above. So assume from now on that u is usc.

We will first treat the special case of u being bounded from above, say $u(z) < M$ for all $z \in \Omega$. The trick is here to consider the functions u_n , $n \in \mathbf{N}$ given by

$$u_n(z) := \sup_{y \in \Omega} \{u(y) - n|y - z|\} \quad \text{for } z \in \Omega$$

u_n is finite valued, since $M \geq u_n(z) \geq u(z_0) - n|z_0 - z|$.

We next prove that u_n is continuous: Given $z \in \Omega$ and $\epsilon > 0$ we choose $y \in \Omega$ such that

$$u_n(z) < u(y) - n|y - z| + \epsilon$$

Then for any $z' \in \Omega$ we have

$$u_n(z) - u_n(z') < \{u(y) - n|y - z| + \epsilon\} - \{u(y) - n|y - z'|\}$$
$$= n\{|z' - y| - |y - z|\} + \epsilon \leq n|z' - z| + \epsilon$$

and since $\epsilon > 0$ is arbitrary we see that

$$u_n(z) - u_n(z') \leq n|z - z'|$$

Interchanging z and $z\prime$ we get

$$\left|u_n(z) - u_n(z')\right| \leq n|z - z'|$$

proving the continuity.

Next we note that

$$u(y) - n|y - z| \geq u(y) - (n + 1)|y - z|$$

which implies that $u_n(z) \geq u_{n+1}(z)$, i.e. that $\{u_n\}$ is a decreasing sequence.

Since

$$u_n(z) = \sup_y \{u(y) - n|y - z|\} \geq u(z) - n|z - z| = u(z)$$

it is left to prove that $\lim_{n \to \infty} u_n(z) \leq u(z)$, i.e. for any prescribed $K > u(z)$ that $u_n(z) \leq K$ for sufficiently large n. For that we use that u is usc : There is a ball $B(z, \delta)$ such that $u(y) < K$ for all $y \in B(z, \delta)$. Now,

$$u_n(z) = \sup_y \{u(y) - n|y - z|\}$$
$$= \max \left\{ \sup_{|y-z|<\delta} \{u(y) - n|y - z|\}, \sup_{|y-z| \geq \delta} \{u(y) - n|y - z|\} \right\}$$
$$\leq \max \{K, M - n\delta\}$$

so if n is so large that $M - n\delta < K$ we have $u_n(z) \leq K$ as desired.

We shall finally deal with the general case in which u is no longer necessarily bounded from above. Choose an increasing homeomorphism $\phi : [-\infty, \infty] \to [-\infty, 0]$ (E.g. $\phi(t) = -\exp(-t)$). The function $\phi \circ u$ is usc and bounded from above, so the construction in the special case treated above yields a sequence $\{v_n\}$ of continuous real valued functions with the property that $v_n(z) \downarrow \phi(u(z))$ as $n \to \infty$ for all $z \in \Omega$. It suffices to verify that $-\infty < v_n(z) < 0$ for all $z \in \Omega$, because we may then choose $u_n := \phi^{-1} \circ v_n$.

The left inequality is trivial because v_n is finite valued. To settle the other inequality we use the definition of v_n, i.e.

$$v_n(z) = \sup_y \{\phi(u(y)) - n|y - z|\}$$

By hypothesis $u(z) < \infty$, so $a := \phi(u(z)) < 0$, and by upper semi-continuity of $\phi \circ u$ there exists $\delta > 0$ such that

$$\phi(u(y)) < \frac{a}{2} < 0 \text{ for all } y \in B(z, \delta)$$

Thus

$$v_n(z) = \max\left\{ \sup_{|y-z|<\delta} \{\phi(u(y)) - n|y - z|\}, \sup_{|y-z|\geq\delta} \{\phi(u(y)) - n|y - z|\} \right\}$$
$$\leq \max\left\{ \frac{a}{2}, 0 - n\delta \right\} < 0$$

$\square$

Section 2 Introductory properties of subharmonic functions

§2. *Introductory properties of subharmonic functions.*

Throughout the remainder of this chapter Ω denotes an open, connected, non-empty subset of the complex plane.

Definition 5.
A map $u : \Omega \to [-\infty, \infty[$ is said to be subharmonic in Ω if
(a) u is upper semi-continuous (usc) and not identically $-\infty$.
(b) For each $z \in \Omega$ there is a ball $B(z, R_z) \subseteq \Omega$ such that

$$u(z) \leq \frac{1}{2\pi} \int_0^{2\pi} u\left(z + re^{i\theta}\right) d\theta \text{ for all } r \in]0, R_z[$$

Remark: The integral in (b) makes sense: Fix z and $r > 0$. Since u is usc, then the function $\theta \to u\left(z + re^{i\theta}\right)$ is Borel measurable and bounded from above. Thus u is essentially a negative function. The value of the integral may a priori be $-\infty$; as we shall see in Proposition 9, it is actually finite. There we shall also see that the inequality (b) holds for all r such that $B(z,r) \subseteq \Omega$, not just for sufficiently small r.

A harmonic function is subharmonic by the mean value property.

Theorem 6 (The strong maximum principle).

Let u be subharmonic in Ω. If there is a $z_0 \in \Omega$ such that $u(z_0) = \sup\limits_{z \in \Omega} u(z)$, then u is a constant.

In particular, if $\overline{\Omega}$ is compact, u subharmonic on Ω, usc in $\overline{\Omega}$ and ≤ 0 on $\partial\Omega$, then $u \leq 0$ throughout Ω.

The maximum principle is one of the crucial properties of subharmonic functions.

Proof: The set $M := \{z \in \Omega | u(z) = u(z_0)\}$ is closed in Ω since u is usc. We will show that M is open as well, so let $z \in M$ be arbitrary. Claim:

$$u\left(z + re^{i\theta}\right) = u(z) \text{ for all } \theta \in \mathbf{R} \text{ and } r \in]0, R_z[$$

We prove the claim by contradiction. If it is false then there are $\theta_0 \in [0, 2\pi[$, $r_0 \in]0, R_z[$ and $\epsilon > 0$ so that

$$u\left(z + r_0 e^{i\theta_0}\right) < u(z) - \epsilon$$

Now, u being usc,

$$u\left(z + r_0 e^{i\theta}\right) < u(z) - \epsilon$$

for all θ in interval $I \subseteq]0, 2\pi[$.

Letting $J :=]0, 2\pi[\setminus I$ we get

$$\frac{1}{2\pi} \int_0^{2\pi} u\left(z + r_0 e^{i\theta}\right) d\theta = \frac{1}{2\pi} \left(\int_I + \int_J \right) u\left(z + r_0 e^{i\theta}\right) d\theta$$

$$\leq \frac{1}{2\pi} \int_I (u(z) - \epsilon) d\theta + \frac{1}{2\pi} \int_J u(z) d\theta$$

$$= u(z) - \epsilon \frac{|I|}{2\pi}$$

which contradicts (b) and hence proves the claim.

The claim shows that u equals $u(z_0)$ in a ball around z so M is open. By connectedness $M = \Omega$. $\qquad\square$

Theorem 7.

Let u be subharmonic in Ω, and let $B[a, R]$ be a closed ball contained in Ω. Then

$$u(z) \leq \int\limits_0^{2\pi} u\left(a + Re^{i\theta}\right) P_R(\theta, z - a) d\theta \text{ for all } z \in B(a, R)$$

where P_R denotes the Poisson kernel.

Proof : Since u is usc it is the pointwise limit of a decreasing sequence $\{u_n\}$ of continuous functions. The function v_n defined by

$$v_n(z) := \begin{cases} \int\limits_0^{2\pi} u_n(a + Re^{i\theta}) P_R(\theta, z - a) d\theta & \text{for } z \in B(a, R) \\ u_n(z) & \text{for } |z - a| = R \end{cases}$$

is by Theorem XII.8 continuous in $B[a, R]$ and harmonic in its interior. $u - v_n$ is usc in $B[a, R]$, subharmonic in its interior and ≤ 0 on its boundary. So by Theorem 6 $u - v_n \leq 0$ throughout $B(a, R)$. The result follows now from the monotone convergence theorem. $\qquad\square$

Theorem 7 explains the terminology subharmonic: The graph of u is below the harmonic function which has the same boundary values on the circle $|z - a| = R$ as u. So subharmonic functions are related to harmonic functions in the same way as convex functions are related to linear functions in one dimension.

Section 3 On the set where $u = -\infty$

In this § we deduce that the set of points in which a subharmonic function takes the value $-\infty$ is so small that all our integrals are finite.

Proposition 8.

A function u which is subharmonic on Ω, is locally integrable in Ω.

In particular, the set $\{z \in \Omega \mid u(z) = -\infty\}$ is a null set, and u cannot be identically $-\infty$ on any non-empty open subset of Ω.

It can be proved that $\{z \in \Omega \mid u(z) = -\infty\}$ is a set of capacity 0 (See [HK;Theorem 5.32]).

Proof : We first observe that u is integrable over $B(a, R)$ if $B[a, R]$ is contained in Ω and $u(a) > -\infty$:

Indeed, multiplying the inequality

$$u(a) \leq \frac{1}{2\pi} \int\limits_0^{2\pi} u\left(a + re^{i\theta}\right) d\theta$$

by πr and integrating from 0 to $R = R_a$ we get

$$-\infty < \frac{1}{2}\pi R^2 u(a) \le \frac{1}{2}\int\limits_0^R \int\limits_0^{2\pi} u\left(a + re^{i\theta}\right)d\theta r\,dr = \int\int\limits_{B(a,R)} u(x,y)dx\,dy$$

As a consequence, if u is not locally integrable at a then u must be identically $-\infty$ throughout some neighborhood of a. Thus the complement in Ω of the set

$$\{z \in \Omega \mid u \text{ is locally integrable at } z\}$$

is the set

$$\{z \in \Omega \mid u \text{ is identically } -\infty \text{ in some neighborhood of } z\}$$

Both these sets are clearly open, so by the connectedness of Ω one of them is empty. But u is not identically $-\infty$ by assumption, so the latter set is empty, and so the first one is Ω. $\square$

The integral in the defining inequality (b) of Definition 5 may a priori have the value $-\infty$. It is, however, finite:

Proposition 9.

Let u be subharmonic on Ω. Let $a \in \Omega$ and $R > 0$ be such that $B[a, R] \subseteq \Omega$, and let I be the function given by

$$I(r) := \frac{1}{2\pi}\int\limits_0^{2\pi} u\left(a + re^{i\theta}\right)d\theta \text{ for } r \in [0, R]$$

Then I is increasing, finite and continuous for $r \in\;]0, R]$, and the map $t \to I\left(e^t\right)$ is convex in the interval $\;]-\infty, \mathrm{Log}\,R]$.

In particular

$$u(a) \le \frac{1}{2\pi}\int\limits_0^{2\pi} u\left(a + re^{i\theta}\right)d\theta \text{ for } r \in [0, R]$$

a result that extends (b) from Definition 5.

Proof: To establish that I is increasing we use the notation of the proof of Theorem 7. We saw that $u \le v_n$. For $r < R$ we have by the mean value property for harmonic functions that

$$\frac{1}{2\pi}\int\limits_0^{2\pi} u_n\left(a + Re^{i\theta}\right)d\theta = v_n(a) = \frac{1}{2\pi}\int\limits_0^{2\pi} v_n\left(a + re^{i\theta}\right)d\theta$$

$$\ge \frac{1}{2\pi}\int\limits_0^{2\pi} u\left(a + re^{i\theta}\right)d\theta$$

and the monotone convergence theorem takes over.

We next show by contradiction that I is finite valued on $]0, R]$. Assume that $I(r_0) = -\infty$ for some $r_0 \in]0, R]$. Then $I(r) = -\infty$ for all $r \in [0, r_0]$, because I is increasing, i.e.

$$\frac{1}{2\pi} \int_0^{2\pi} u\left(a + re^{i\theta}\right) d\theta = -\infty \text{ for all } r \in [0, r_0]$$

This implies (integrate with respect to r from 0 to r_0) that

$$\int_{B(a, r_0)} u(x, y) dx dy = -\infty$$

contradicting the local integrability (Proposition 8).

A convex function on an open interval is continuous, so left is only the convexity statement. We postpone a proof of that to the next section where we will possess more machinery. $\qquad\Box$

Section 4 Approximation by smooth functions

The following characterization of smooth subharmonic functions is very useful and connects once more the concepts of subharmonic and harmonic functions.

Theorem 10.
Let $u \in C^2(\Omega, \mathbf{R})$. Then u is subharmonic on Ω if and only if $-\Delta u \leq 0$ on Ω.

Proof:
Suppose first that $\Delta u \geq 0$. We shall for $z \in \Omega$ verify that

$$u(z) \leq \frac{1}{2\pi} \int_0^{2\pi} u\left(z + re^{i\theta}\right) d\theta \text{ for all sufficiently small } r > 0$$

Let us for simplicity take $z = 0$.

The expression for the Laplacian Δ in polar coordinates (r, θ) is

$$(1) \quad \Delta u = u_{rr} + \frac{1}{r} u_r + \frac{1}{r^2} u_{\theta\theta}$$

When we integrate (1) with respect to θ from 0 to 2π (noting that the last term drops out because $\phi \to u_\theta(re^{i\phi})$ is periodic in ϕ) we get with the notation

$$I(r) := \frac{1}{2\pi} \int_0^{2\pi} u\left(re^{i\theta}\right) d\theta$$

205

that

$$(2) \quad \frac{1}{2\pi} \int\limits_0^{2\pi} \Delta u\left(re^{i\theta}\right) d\theta = I'' + \frac{1}{r}I' = \frac{1}{r}(rI')'$$

Combining (2) with $\Delta u \geq 0$ we see that $r \to rI'(r)$ is an increasing function. Since $I'(r)$ clearly is bounded (for small r), $rI'(r) \to 0$ as $r \to 0$. But then $rI'(r) \geq 0$ so that $I'(r) \geq 0$ and thus I is an increasing function of r. Now, as desired

$$u(0) = I(0) \leq I(r) = \frac{1}{2\pi} \int\limits_0^{2\pi} u\left(re^{i\theta}\right) d\theta$$

Let next u be subharmonic. We assume that $-\Delta u(z) > 0$ for some $z \in \Omega$ and shall derive a contradiction. Take again $z = 0$ for the sake of simplicity.

Applying the first part of the proof to $-u$ we get that I is decreasing for small r. But u is subharmonic so I is actually increasing (Proposition 9); hence I is constant and the right hand side of (2) is zero. But that contradicts the assumption $-\Delta u(0) > 0$. $\square$

A subharmonic function need not be C^2. But it can be approximated by smooth subharmonic functions:

Theorem 11.

Let u be subharmonic in Ω. Let ω be an open subset of Ω with the property that $dist(\omega, \partial\Omega) > 0$.

Then there is a decreasing sequence $u_1 \geq u_2 \geq \cdots$ of subharmonic functions in $C^\infty(\omega)$ such that $u_n(z) \underset{n \to \infty}{\to} u(z)$ for all $z \in \Omega$.

Proof : The standard mollifying method (convolution) works. It goes as follows:
Let $0 < R < \frac{1}{2}\text{dist}(\omega, \partial\Omega)$. Then the function

$$U(z) := \begin{cases} u(z) & \text{when } \text{dist}(z, \omega) < R \\ 0 & \text{otherwise on } \mathbf{R}^2 \end{cases}$$

is locally integrable in $\mathbf{R}^2$, since u is locally integrable in Ω. Let $\phi \in C_0^\infty(\mathbf{R}^2)$ be a positive radial function such that $supp\,\phi \subseteq B[0, R]$ and $\int_{\mathbf{R}^2} \phi = 1$ and define for $n = 1, 2, \cdots$ the functions

$$\phi_n(z) := n^2\phi(nz) \text{ for } z \in \mathbf{R}^2$$

Then

$$\phi_n \in C_0^\infty(\mathbf{R}^2) \,, \ supp\,\phi_n \subseteq B\left[0, \frac{R}{n}\right] \text{ and } \int_{\mathbf{R}^2} \phi_n = 1$$

Let U_n be the convolution of U and ϕ_n, i.e. the function given by the expression (the integration is over $\mathbf{R}^2$):

$$U_n(z) := \int U(y)\phi_n(z-y)dy \text{ for } z \in \mathbf{R}^2$$

Clearly $U_n \in C^\infty(\mathbf{R}^2)$, because we may differentiate under the integral sign.

Let us next prove that U_n is subharmonic on ω. Let $z \in \Omega$ and let $0 < r \le R$. Then (using Proposition 9 and that ϕ is radial)

$$U_n(z) = \int U(y)\phi_n(z-y)dy = \int U(z+y)\phi_n(y)dy$$

$$= \int_{B(0,\frac{R}{n})} U(z+y)\phi_n(y)dy = \int_{B(0,\frac{R}{n})} u(z+y)\phi_n(y)dy$$

$$\le \int_{B(0,\frac{R}{n})} \left\{ \frac{1}{2\pi} \int_0^{2\pi} u\left(z+y+re^{i\theta}\right)d\theta \right\} \phi_n(y)dy$$

$$= \frac{1}{2\pi} \int_0^{2\pi} \int_{B(0,\frac{R}{n})} u\left(z+y+re^{i\theta}\right)\phi_n(y)dyd\theta$$

$$= \frac{1}{2\pi} \int_0^{2\pi} U_n\left(z+re^{i\theta}\right)d\theta$$

showing that U_n is subharmonic on ω.

We continue by proving that $U_n(z)$ is decreasing in n when $z \in \omega$:

$$U_n(z) = \int_{B(0,\frac{R}{n})} u(z+y)\phi_n(y)dy = \int_{B(0,\frac{R}{n})} u\left(z+\frac{y}{n}\right)\phi(y)dy$$

$$= \frac{1}{2} \int_0^R \int_0^{2\pi} u\left(z+\frac{\rho e^{i\theta}}{n}\right)d\theta\phi(\rho)\rho d\rho$$

The inner integral is increasing with ρ (Proposition 9), so we get

$$U_n(z) \ge \frac{1}{2} \int_0^R \int_0^{2\pi} u\left(z+\frac{\rho e^{i\theta}}{n+1}\right)d\theta\phi(\rho)\rho d\rho = U_{n+1}(z)$$

The computation just made also reveals that

$$U_n(z) \ge \frac{1}{2} \int_0^R 2\pi u(z)\phi(\rho)\rho d\rho = u(z)$$

so now it is only left to show that

$$\lim_{n \to \infty} U_n(z) \leq u(z) \text{ for all } z \in \omega$$

i.e. given $K > u(z)$ there exists an n so that $U_n(z) \leq K$.

By the usc $u < K$ in a neighborhood V of z. When n is so large that $B(z, R/n) \subseteq V$ we have

$$U_n(z) = \int_{B\left(0, \frac{R}{n}\right)} u(z + y)\phi_n(y)dy \leq \int_{B\left(0, \frac{R}{n}\right)} K\phi_n(y)dy = K$$

□

We shall now keep a promise:

Proof of Proposition 9 (continued):

For simplicity of exposition we take $a = 0$ as usual. Assume first that $u \in C^\infty$. We shall show that the function

$$h(t) := \frac{1}{2\pi} \int_0^{2\pi} u\left(e^t e^{i\theta}\right) d\theta$$

is convex for $-\infty < t < \text{Log } R$.

During the proof of the first part of Theorem 10 we observed that $r \to rI'(r)$ is an increasing function. Hence so is

$$t \to e^t I'\left(e^t\right) = \frac{d}{dt}\left(I(e^t)\right) = h'(t)$$

implying that h is convex.

If u is no longer assumed C^∞ we approximate by a sequence of C^∞ subharmonic functions $u_n \downarrow u$. Then

$$h_n(t) := \frac{1}{2\pi} \int_0^{2\pi} u_n\left(e^t e^{i\theta}\right) d\theta$$

is a decreasing sequence of convex functions, so its limit, which is h, is convex as well. □

Section 5 Constructing subharmonic functions

Proposition 12.

(α) *If u_1 and u_2 are subharmonic in Ω then so are $u_1 + u_2$ and $\max\{u_1, u_2\}$.*

(β) *If u is subharmonic in Ω then so is λu for any $\lambda \in [0, \infty[$.*

Section 5.   Constructing subharmonic functions

(γ) *If $\{u_n\}$ is a decreasing sequence of subharmonic functions in Ω then $\lim u_n$ is subharmonic in Ω, unless it is identically $-\infty$.*

Proof : Left to the reader.   □

For the next result we need (a particular case of) Jensen's inequality for integrals.

Theorem 13 (Jensen's inequality for integrals).
If f is a real valued function in $L^1(0, 2\pi)$ such that $a < f(t) < b$ and if ϕ is convex on $]a, b[]a,b[$, then

$$\phi\left\{\int\limits_0^{2\pi} f(\theta)\frac{d\theta}{2\pi}\right\} \le \int\limits_0^{2\pi} (\phi \circ f)(\theta)\frac{d\theta}{2\pi}$$

The cases $a = -\infty$ and $b = \infty$ are allowed.

For a proof consult [Ru;Theorem 3.3].   □

Theorem 14.
(α) *If u is harmonic in Ω and ϕ is convex on the range of u, then $\phi \circ u$ is subharmonic in Ω.*

(β) *If u is subharmonic in Ω and ϕ is a convex increasing function on the range of u, then $\phi \circ u$ is subharmonic in Ω. (We put $\phi(-\infty) = \lim\limits_{t \to -\infty} \phi(t)$).*

Proof :

(α) A convex function on the real line is automatically continuous, so $\phi \circ u$ is continuous. From the mean value property

$$u(a) = \int\limits_0^{2\pi} u\left(a + re^{i\theta}\right)\frac{d\theta}{2\pi}$$

we deduce via Jensen's inequality for integrals that

$$(\phi \circ u)(a) \le \int\limits_0^{2\pi} (\phi \circ u)\left(a + re^{i\theta}\right)\frac{d\theta}{2\pi}$$

so $\phi \circ u$ is subharmonic.

(β) Similar arguments to the ones from (α) work here.   □

Theoretical Examples 15.
(α) *Let $f \in Hol(\Omega)$. If f is not identically 0 then $Log|f|$ and $Log^+|f| := \max\{Log|f|, 0\}$ are subharmonic in Ω.*

209

(β) If f and g are holomorphic in Ω then $|f|^p|g|^q$ is subharmonic for all $p, q \geq 0$.

(γ) If u is harmonic in Ω then $|u|^p$ is subharmonic in Ω for all $p \geq 1$.

Proof :

(α) $\text{Log}|f|$ is clearly usc and not identically $-\infty$. If $f(z) = 0$ at a point $z \in \Omega$ then $\text{Log}|f(z)| = -\infty$, so the inequality (b) from Definition 5 is obviously satisfied at z. If $f(z) \neq 0$ then $\text{Log}|f| = \Re(\text{Log} f)$ is harmonic around z, hence subharmonic.

$\text{Log}^+|f|$ is the composition of $\text{Log}|f|$ with the convex increasing function $t \to t1_{]0,\infty[}(t)$.

(β) Assume first that neither f nor g is identically 0. Then $\text{Log}|f|$ and $\text{Log}|g|$ are subharmonic, hence so is $p\text{Log}|f| + q\text{Log}|g|$. Left is just to compose with the exponential map. The modifications in the cases where f and/or g are identically 0 are left to the reader.

(γ) Theorem 14(α). □

Practical examples are now numerous, e.g. $u(x, y) = |x|$ or more generally $u(x, y) = |x|^p$ for $p \geq 1$.

Section 6 Applications

We present three applications: One is a proof of Radó's theorem, the other is a treatment of H^p−spaces, and the third is a proof of the F. and R. Nevanlinna theorem.

Radó's theorem

Our first application of the theory of subharmonic functions is a proof of Radó's theorem.

Theorem 16 (Radó's Theorem).

If $f \in C(\Omega)$ is holomorphic in $\{z \in \Omega \mid f(z) \neq 0\}$ then f is holomorphic in all of Ω.

Proof : We may assume that f is not identically 0. We write $f = u + iv$, where u and v are real valued.

The function $s := \text{Log}|f|$ is subharmonic in Ω (same proof as in (α) of the Theoretical Examples 15 above). So is $u + \epsilon s$ for any $\epsilon > 0$ (same proof).

We will show that f is holomorphic in any ball in Ω; for simplicity of notation we assume that the ball is $B(0, 1)$ and that $B[0, 1] \subseteq \Omega$. We assume furthermore that $|f(z)| \leq 1$ for $z \in B[0, 1]$. Note that then $s \leq 0$ on $B(0, 1)$.

Now, for any $z \in B[0,1]$ we have by Theorem 7 that

$$u(z) + \epsilon s(z) \le \frac{1}{2\pi} \int_0^{2\pi} \left\{ u\left(e^{i\theta}\right) + \epsilon s\left(e^{i\theta}\right) \right\} P_1(\theta, z) d\theta$$

$$\le \frac{1}{2\pi} \int_0^{2\pi} u\left(e^{i\theta}\right) P_1(\theta, z) d\theta$$

At those points $z \in B(0,1)$ where $f(z) \ne 0$ we get by letting ϵ decrease to 0 that

$$u(z) \le \frac{1}{2\pi} \int_0^{2\pi} u\left(e^{i\theta}\right) P_1(\theta, z) d\theta$$

However, u is continuous and the points where $f(z) \ne 0$ are dense in $B(0,1)$ (Proposition 10), so the inequality holds for all points $z \in B(0,1)$.

Our arguments work in the same way for $-f$, i.e. with $-u$ instead of u. Thus

$$u(z) = \frac{1}{2\pi} \int_0^{2\pi} u\left(e^{i\theta}\right) P_1(\theta, z) d\theta$$

which shows (Theorem XII.8) that u is harmonic. In particular $u \in C^\infty(B(0,1))$.

Replacing f by if we get $v \in C^\infty(B(0,1))$, so $f \in C^\infty(B(0,1))$. f satisfies the Cauchy-Riemann equations on the dense set $\{z \in B(0,1) \mid f(z) \ne 0\}$, and hence on all of $B(0,1)$. So f is holomorphic. $\qquad\square$

Hardy Spaces

Definition 17.

Let f be holomorphic in the unit disc. f is said to belong to the Hardy space H^p , $p \in]0, \infty[$, if

$$\sup_{0 < r < 1} \frac{1}{2\pi} \int_0^{2\pi} \left| f\left(re^{i\theta}\right) \right|^p d\theta < \infty$$

We leave it to the reader to verify that H^p is a linear subspace of $Hol(B(0,1))$.

Theorem 18.

If $f \in H^p$ where $p \ge 1$ then there exists exactly one $f^* \in L^p(S^1)$ such that

$$f(z) = \frac{1}{2\pi} \int_0^{2\pi} P_1(t, z) f^*(e^{it}) dt \text{ for all } |z| < 1$$

Proof: We prove the theorem for $p = 1$; for $p > 1$ the argument is similar, but simpler.

Let $0 < r_1 < r_2 < \cdots < 1$ be an increasing sequence such that $r_n \to 1$ as $n \to \infty$, and define

$$g_n(e^{it}) := |f(r_n e^{it})|^{\frac{1}{2}} \text{ for } t \in \mathbf{R}$$

That $f \in H^1$ means that $\{g_n\}$ is a bounded subset of $L^2(S^1)$. Since the unit ball of L^2 is weakly sequentially compact (See [DS; Corollary IV.8.4]), we may (possibly by taking a subsequence) assume that there exists a function $g \in L^2(S^1)$ such that $g_n \to g$ weakly in L^2. Recalling that $|f|^{\frac{1}{2}}$ is subharmonic (Theoretical Examples 15(β)) we get by Proposition 9 for $m < n$ that

$$g_m(e^{it}) \leq \int_0^{2\pi} g_n(e^{i\theta}) P_{r_n}(\theta, z) d\theta \quad \text{where } z = r_m e^{it}$$

and letting $n \to \infty$ it follows that

$$g_m(e^{it}) \leq \int_0^{2\pi} g(e^{i\theta}) P_1(\theta, z) d\theta$$

By the Cauchy-Schwarz inequality we get

$$g_m(e^{it})^2 \leq \int_0^{2\pi} \left\{ g(e^{i\theta}) \left(P_1(\theta, z)^{\frac{1}{2}} \right) \right\}^2 d\theta \int_0^{2\pi} P_1(\theta, z) d\theta$$

$$= \int_0^{2\pi} g(e^{i\theta})^2 P_1(\theta, z) d\theta$$

which we integrate with respect to t to get $\|g_m\|_2^2 \leq \|g\|_2^2$.

This inequality combined with the fact that g_m converges weakly to g implies that $g_m \to g$ in L^2. And then $g_m^2 \to g^2$ in L^1. Indeed,

$$\int |g_m^2 - g^2| dt = \int |g_m - g| |g_m - g| dt$$

$$\leq \|g_m - g\|_2 \|g_m + g\|_2 \leq 2 \|g_m - g\|_2 \|g\|_2$$

In particular the sequence $\{g_n^2 = |f(r_n e^{it})|\}$ is uniformly integrable. Hence so is the sequence $\{t \to f(r_n e^{it})\}$. By the Dunford-Pettis compactness criterion (See [DS]) we may (possibly by taking a subsequence) assume that there exists an $f^* \in L^1(S^1)$ such that $f(r_n e^{it}) \to f^*(e^{it})$ weakly in L^1.

Letting $n \to \infty$ in the Poisson formula

$$f(z) = \frac{1}{2\pi} \int_0^{2\pi} f(r_n e^{it}) \frac{r_n^2 - |z|^2}{|r_n e^{it} - z|^2} dt$$

we get the existence part of the theorem.

The uniqueness will follow once we check for any L^1-function f^* that

$$\frac{1}{2\pi} \int_0^{2\pi} P_1(t,z) f^*(e^{it})\, dt = 0 \text{ for all } |z| < 1$$

implies $f^* = 0$. It suffices to prove the statement for f^* real valued. In that case we get by property (iii) of the Poisson kernel that

$$0 = \frac{1}{2\pi} \int_0^{2\pi} P_1(t,z) f^*(e^{it})\, dt$$

$$= \Re\left\{ \frac{1}{2\pi} \int_0^{2\pi} f^*(e^{it})\, dt + 2 \sum_{n=1}^{\infty} \frac{1}{2\pi} \int_0^{2\pi} e^{-int} f^*(e^{it})\, dt\, z^n \right\}$$

The parenthesis is an analytic function with real part 0, so it is a purely imaginary constant c, i.e.

$$c = \frac{1}{2\pi} \int_0^{2\pi} f^*(e^{it})\, dt + 2 \sum_{n=1}^{\infty} \frac{1}{2\pi} \int_0^{2\pi} e^{-int} f^*(e^{it})\, dt\, z^n$$

Taking $z = 0$ we see that $c = 0$, because f^* is real. Now,

$$\frac{1}{2\pi} \int_0^{2\pi} e^{-int} f^*(e^{it})\, dt = 0 \text{ for all } n \geq 0$$

Since f^* is real, complex conjugation shows that this holds for all $n \in \mathbf{N}$, i.e. that all the Fourier coefficients of f^* are 0. But then $f^* = 0$. $\square$

The following corollary which is equivalent to the theorem just proved, has been generalized extensively and is useful in prediction theory.

Theorem 19 (F. and M. Riesz's theorem).
Let μ be a complex Borel measure on S^1 for which

$$\int_{S^1} z^n d\mu(z) = 0 \text{ for all } n = 1, 2, \cdots$$

Then μ is absolutely continuous with respect to the linear measure on S^1.

Proof : Expanding $P_1(t,z)$ in powers of z and $\bar{z}$ and using the condition on μ we see that

$$f(z) := \int_0^{2\pi} P_1(t,z) d\mu(e^{it}) \text{ for } |z| < 1$$

213

is a power series in z alone, so that f is holomorphic in $B(0,1)$. Using Fubini we find that $f \in H^1$.

When we apply Theorem 18 and once again expand P_1 in z and $\overline{z}$ we find that

$$\frac{1}{2\pi} \int_0^{2\pi} e^{int} d\mu(t) = \frac{1}{2\pi} \int_0^{2\pi} e^{int} f^*(e^{it}) dt \quad \text{for all } n \in \mathbf{Z}$$

Since the trigonometric polynomials are dense in the periodic continuous functions we get for all $g \in C(S^1)$ that

$$\frac{1}{2\pi} \int_0^{2\pi} g(t) d\mu(t) = \frac{1}{2\pi} \int_0^{2\pi} g(t) f^*(e^{it}) dt$$

which finishes the proof. $\qquad\qquad\qquad\qquad\qquad\qquad\qquad\qquad\qquad\square$

The f^* of Theorem 18 above cannot be an arbitrary function in $L^1(S^1)$ as the following result shows:

Theorem 20.

Let f and f^* be as in Theorem 18. Assume that f is not identically 0. Then

$$Log|f(0)| \le \frac{1}{2\pi} \int_0^{2\pi} Log\left|f^*\left(e^{i\theta}\right)\right| d\theta$$

and f^* cannot vanish on any set of positive linear measure on S^1.

Proof : In the course of proving Theorem 18 we showed that the sequence $\left\{\left|f(r_n e^{i\theta})\right|\right\}$ converges to $\left|f^*(e^{i\theta})\right|$ in L^1. Choosing a subsequence if necessary we may assume that

$$\left|f\left(r_n e^{i\theta}\right)\right| \xrightarrow[n \to \infty]{} \left|f^*\left(e^{i\theta}\right)\right| \quad \text{for almost all } \theta$$

Since

$$\left|f\left(r_n e^{i\theta}\right)\right| - Log\left|f\left(r_n e^{i\theta}\right)\right| \ge 0$$

Fatou's lemma implies that

$$\limsup_{n \to \infty} \int_0^{2\pi} Log\left|f\left(r_n e^{i\theta}\right)\right| d\theta \le \int_0^{2\pi} Log\left|f^*\left(e^{i\theta}\right)\right| d\theta$$

By Jensen's inequality,

$$Log|f(0)| \le \frac{1}{2\pi} \int_0^{2\pi} Log\left|f\left(r_n e^{i\theta}\right)\right| d\theta$$

finishing the proof of the first statement.

The last statement is a triviality if $f(0) \neq 0$. If f has a zero of order N at $z = 0$ we apply the above result to $F(z) := f(z)z^{-N}$, noting that $F^*(e^{it}) = f^*(e^{it})e^{-iNt}$. $\square$

If $f \in H^p$ then $f \in Hol(D) \cap L^p(D)$ for all $p \in [1, \infty[$ (Use polar coordinates). The converse is not true:

Example 21.

The function $f(z) := (1 - z)^{-1}$ belongs to $Hol(D) \cap L^1(D)$, but not to H^1:

Clearly $f \in Hol(D)$, and we leave it to the reader to check that $f \in L^p(D)$ for all $p \in [1, 2[$.

We assume now that $f \in H^1$ and derive a contradiction from that assumption: During the proof of Theorem 18 we saw that $f(r_n e^{it}) \underset{n \to \infty}{\to} f^*(e^{it})$ weakly in L^1. It follows that

$$f^*(e^{it}) = \frac{1}{1 - e^{it}} \ , \ so \ \left(t \to \frac{1}{1 - e^{it}}\right) \in L^1(-\pi, \pi)$$

But

$$\left|1 - e^{it}\right| = \left|\sum_{n=1}^{\infty} \frac{(it)^n}{n!}\right| = |t| \left|\sum_{n=0}^{\infty} \frac{(it)^n}{(n+1)!}\right| \leq 2|t|$$

for small t, so (again for t small)

$$\left|\frac{1}{1 - e^{it}}\right| \geq \frac{1}{2|t|}$$

and the right hand side is not integrable over any open interval around 0. $\square$

F. and R. Nevanlinna's Theorem

Theorem 22 (F. and R. Nevanlinna's Theorem).

Let f be analytic in the unit disc. Then f is the quotient between two bounded analytic functions, iff

$$\sup_{0 < r < 1} \left\{\frac{1}{2\pi} \int_0^{2\pi} Log^+ |f(re^{it})| dt\right\} < \infty$$

Proof : Let us first assume that

$$(*) \quad \sup_{0 < r < 1} \left\{\frac{1}{2\pi} \int_0^{2\pi} w(re^{it}) dt\right\} < \infty \ , \ \text{where } w := Log^+ |f|$$

Note that w is subharmonic (Example 15).

The harmonic function u_r on $B(0, r)$ that on the boundary coincides with w, increases with r (Remark immediately after Theorem 7). The values at 0 of the functions u_r are by the mean value property (Theorem XII.6) bounded by (*), so according to Harnack's theorem (Theorem XII.10) $u_r \to u$ as $r \to 1_-$, where u is harmonic on $B(0, 1)$ and nonnegative, because $w \leq u$.

Choose a holomorphic function g such that $u = \Re g$ (Theorem XII.2(β)). Then $|h| = e^u \geq 1$ and $|f| \leq |h|$, so both $1/h$ and f/h are analytic and bounded by 1. We are through, since their quotient is f.

Suppose conversely that $f = a/b$ where a and b are bounded analytic functions. We may assume that $|a| \leq 1$, $|b| \leq 1$ and that $b(0) \neq 0$. Then

$$Log^+ |f| \leq -Log\,|b|$$

and by the subharmonicity of $Log\,|b|$ (Example 15)

$$-\infty < Log\,|b(0)| \leq \frac{1}{2\pi} \int_0^{2\pi} Log\,\left|b\left(re^{i\theta}\right)\right| d\theta$$

Thus

$$\frac{1}{2\pi} \int_0^{2\pi} Log^+ \left|f\left(re^{i\theta}\right)\right| d\theta \leq -\frac{1}{2\pi} \int_0^{2\pi} Log\left|b\left(re^{i\theta}\right)\right| d\theta \leq -Log\,|b(0)|$$

and the right hand side is independent of r. $\square$

Section 7 Exercises

1. For which values of $a, b, c \in \mathbf{R}$ is the polynomial

$$u(x, y) := ax^2 + 2bxy + cy^2$$

a harmonic function? a subharmonic function?

2. Let $\phi : \Omega \to \Omega_0$ be a holomorphic map between two domains in the complex plane. Let u be subharmonic on Ω_0.

Show that $u \circ \phi$ is subharmonic on Ω unless it is identically $-\infty$. Hint: $\Delta(u \circ \phi) = |\phi'|^2 \Delta u$, when ϕ is holomorphic.

3. Let K be a compact subset of Ω. Let u be subharmonic on Ω. Let h be a real valued continuous function on K such that h is harmonic on $int(K)$ and $u \leq h$ on ∂K.

Show that $u \leq h$ on all of K.

4. Let $f \in Hol(B(0, 1))$ have the power series expansion

$$f(z) = \sum_{n=0}^{\infty} a_n z^n \quad \text{for } |z| < 1$$

Show that

$$f \in H^2 \quad \Leftrightarrow \quad \sum_{n=0}^{\infty} |a_n|^2 < \infty$$

5. Let $f \in Hol(B(0,1))$ and let $p \in [1, \infty]$.

(α) Let $f \in H^p$. Show that there exists a harmonic function u such that $|f|^p \leq u$.
Hint: Theorem XIII.20.

(β) Assume that there exists a harmonic function u such that $|f|^p \leq u$. Show that $f \in H^p$.

[This exercise gives us a hint as to how to define the Hardy space $H^p(\Omega)$ when Ω is a domain in $\mathbf{C}$: $f \in Hol(\Omega)$ is in $H^p(\Omega)$ if there exists a harmonic function u on Ω such that $|f|^p \leq u$.

6. Let u be a convex function on an open convex subset of $\mathbf{R}^2$. Show that u is subharmonic. *Hint:* A convex function on an open set is automatically continuous, so

$$\frac{1}{2\pi} \int_0^{2\pi} u\left(a + re^{i\theta}\right) d\theta$$

makes sense as a Riemann integral.

7. Let u be usc, and let f be continuous and increasing on the range of u. Show that $f \circ u$ is usc.

8. Show that a continuous function with the mean value property is harmonic. Hint: Theorem 7.

9. Prove

Schwarz's reflection principle.

Let Ω be an open subset of $\mathbf{R}^2$ such that Ω is symmetric around the x-axis. Let u be harmonic on $\{(x,y) \in \Omega \,|\, y > 0\}$, continuous on $\{(x,y) \in \Omega \,|\, y \geq 0\}$ and 0 on $\{(x,y) \in \Omega \,|\, y = 0\}$.

Then there exists a harmonic function U on Ω such that $U = u$ on $\{(x,y) \in \Omega \,|\, y > 0\}$.

Hint: Show that

$$U(x,y) := \begin{cases} u(x,y) & \text{for } y \geq 0 \\ -u(x,-y) & \text{for } y < 0 \end{cases}$$

has the mean value property.

10. Let u be subharmonic on an open subset Ω of $\mathbf{R}^2$. Let $B[0,r] \subseteq \Omega$, where $r > 0$. Define v on Ω by

$$v(z) := \begin{cases} \frac{1}{2\pi} \int_0^{2\pi} u\left(re^{i\theta}\right) P_r(\theta, z) d\theta & \text{for } z \in B(0,r) \\ u(z) & \text{for } z \in \Omega \backslash B(0,r) \end{cases}$$

Show that v is subharmonic on Ω.

11. We have seen that $|h|$ is subharmonic if h is harmonic. Show the following more general statement:

If $\alpha \geq 1$ and $h_1, h_2, \cdots, h_n$ are harmonic, then

$$\left(|h_1|^{\alpha} + |h_2|^{\alpha} + \cdots + |h_n|^{\alpha}\right)^{\frac{1}{\alpha}}$$

is subharmonic.

12. Use the procedure from the proof of Radó's theorem to show the theorem on an isolated singularity for a harmonic function (Theorem XII.5).

Chapter 14 Various Applications

Section 1 The Phragmén-Lindelöf principle

The weak maximum principle (Corollary IV.16), which says that an analytic function never exceeds its maximal boundary value, does not hold in general for unbounded domains. For example, if Ω is the lower half plane, so that $\partial\Omega$ is the real axis, then $f(z) := \exp(iz)$ is bounded on $\partial\Omega$ but not on Ω. So we need extra restrictions on the function and the domain to get maximum principles for unbounded domains. The so-called Phragmén-Lindelöf principle is a way of finding such maximum principles. Our version is the following.

Theorem 1 (Phragmén-Lindelöf's principle).

Let Ω be a connected open subset of $\mathbf{C}$ and let ϕ be a function in $C(\overline{\Omega}) \cap Hol(\Omega)$ which is bounded and not identically 0.

If $f \in C(\overline{\Omega}) \cap Hol(\Omega)$ and the constant $M \geq 0$ satisfy that

(1) $|f(z)| \leq M$ for all $z \in \partial\Omega$, and

(2) $|f(z)||\phi(z)|^\eta \to 0$ as $z \to \infty$ in Ω for each fixed $\eta > 0$,

then $|f(z)| \leq M$ for all $z \in \Omega$.

Proof: The case $M = 0$ is a consequence of the case $M > 0$, so we may assume that $M > 0$. Let $B > 0$ be a bound of $|\phi|$. Consider for $\eta > 0$ the continuous function

$$u_\eta(z) := |f(z)|\left(\frac{|\phi(z)|}{B}\right)^\eta \text{ for } z \in \overline{\Omega}$$

which is subharmonic on Ω (Example XII.17(β)). By the assumption (2) we may choose $R > 0$ so large that

$$u_\eta(z) \leq M \text{ for all } z \in \Omega \text{ such that } |z| \geq R$$

Since $\partial(\Omega \cap B(0,R)) \subseteq \partial\Omega \cup \{z \in \Omega \mid |z| = R\}$ we get by the maximum principle for subharmonic functions (Theorem XII.6) that

$$u_\eta(z) \leq \max\{M, M\} = M \text{ for } z \in \Omega \cap B(0,R)$$

This inequality is also true outside of the ball $B(0,R)$ by our choice of R, so $u_\eta(z) \leq M$ for $z \in \Omega$, i.e.

$$|f(z)|\left(\frac{|\phi(z)|}{B}\right)^\eta \leq M \text{ for } z \in \Omega$$

Letting $\eta \to 0$ for fixed $z \in \Omega$ we get that $|f(z)| \leq M$ for all those $z \in \Omega$ for which $\phi(z) \neq 0$. However, the zeros of ϕ are isolated in Ω (Theorem IV.11), so the desired inequality holds by the continuity of f everywhere in Ω. $\qquad\square$

The connectivity assumption was only used at the very end of the proof. We may delete that assumption if we in return require that ϕ vanishes nowhere.

Corollary 2.

Let $\Omega \neq \mathbb{C}$ be an open subset of the complex plane. Let $f \in C\left(\overline{\Omega}\right) \cap Hol(\Omega)$ be bounded.

If $|f| \leq M$ on $\partial\Omega$, then $|f| \leq M$ on all of Ω.

Proof: We may of course assume that $\Omega \neq \emptyset$, so that also $\partial\Omega \neq \emptyset$. Let $\epsilon > 0$. By continuity of f there exists a point $z_0 \in \Omega$ such that $|f(z_0)| < M + \epsilon$ and then there is even a ball $B[z_0, r] \subseteq \Omega$ such that $|f| \leq M + \epsilon$ on that ball. Applying Theorem 1 to the set $\Omega \backslash B[z_0, r]$ and the function $\phi(z) := (z - z_0)^{-1}$ we get that

$$|f(z)| \leq M + \epsilon \text{ for all } z \in \Omega \backslash B[z_0, r]$$

This inequality also holds for $z \in B[z_0, r]$ by the very construction of that ball, so $|f(z)| \leq M + \epsilon$ for all $z \in \Omega$. But $\epsilon > 0$ was arbitrary. □

Corollary 3 (The three lines theorem).

Let $\Omega := \{z \in \mathbb{C} \mid 0 < \Re z < 1\}$. Let f be bounded and holomorphic on Ω and continuous on the closure $\overline{\Omega}$ of Ω and put

$$(1) \quad M(x) := \sup_{-\infty < y < \infty} \{|f(x + iy)|\}$$

Then $\log M$ is a convex function on $[0, 1]$. In particular

$$(2) \quad M(x) \leq M(0)^{1-x} M(1)^x \text{ for all } x \in [0, 1]$$

Proof: Let $g(z) := e^{az} f(z)$, where a is a real constant. Applying Corollary 2 to g we see for any $x \in]0, 1[$ that

$$|g(x + iy)| \leq \max \{M(0), M(1)e^a\}$$

and so that

$$(3) \quad M(x) \leq \max \left\{ M(0)e^{-ax}, M(1)e^{a(1-x)} \right\}$$

If $M(0) = 0$ then (3) says that $M(x) \leq M(1)e^{a(1-x)}$. Since this holds for all real numbers a, we see that $M(x) = 0$, and so (2) is trivially satisfied. The same kind of arguments works if $M(1) = 0$.

If both $M(0)$ and $M(1)$ are different from 0 we choose a so that $M(0) = M(1)e^a$ (this minimizes the right hand side of the inequality (3)); then (3) simplifies as desired to

$$M(x) \leq e^{-ax} M(0) = M(0)^{1-x} M(1)^x$$

If we apply the same arguments to the strip

$$\{z \in \mathbf{C} \mid s \leq \Re z \leq t\} \quad \text{where} \quad 0 \leq s < t \leq 1$$

as we did to the entire strip $\{z \mid 0 \leq \Re z \leq 1\}$ we find that

$$M(x)^{t-s} \leq M(s)^{t-x} M(t)^{x-s}$$

i.e. that $\log M$ is convex.   □

Section 2 The Riesz-Thorin interpolation theorem

Using The Three Lines Theorem we shall now prove the M.Riesz-Thorin interpolation theorem. The interpolation theorem holds for general measure spaces, but for simplicity we will only consider $\mathbf{R}^d$ with Lebesgue measure. A proof for the general case can be found in [DS;VI.10].

We need some preliminaries:

A simple function is a (finite) linear combination of indicator functions for measurable sets of finite measure. We let S denote the complex vector space of simple functions.

If $1 \leq q_0 < q_1 \leq \infty$ and $f \in L^{q_0} \cap L^{q_1}$, then $f \in L^r$ for each $r \in [q_0, q_1]$:
Indeed, if $q_1 < \infty$, then

$$|f|^r \leq |f|^{q_0} 1_A + |f|^{q_1} 1_B$$

where

$$A := \left\{ x \in \mathbf{R}^d \mid |f(x)| \leq 1 \right\} \quad \text{and} \quad B := \left\{ x \in \mathbf{R}^d \mid |f(x)| > 1 \right\}$$

If $q_1 = \infty$, then we may scale f down so that we may assume that $|f| \leq 1$; and then $|f|^r \leq |f|^{q_0}$, so we are through also in this case.

Theorem 4 (Riesz-Thorin's interpolation theorem).

Let $1 \leq p_0, q_0, p_1, q_1 \leq \infty$, and let $T : S \to L^{q_0} \cap L^{q_1}$ be a linear operator such that the mappings $T : S \to L^{q_0}$ and $T : S \to L^{q_1}$ have finite norms M_0 and M_1 , when S is equipped with the norms from L^{p_0} and L^{p_1} respectively.

For $t \in [0,1]$ we define p_t and q_t by the formulas

$$\frac{1}{p_t} = \frac{1-t}{p_0} + \frac{t}{p_1} \quad \text{and} \quad \frac{1}{q_t} = \frac{1-t}{q_0} + \frac{t}{q_1}$$

Then T maps S into L^{q_t}, and its norm is at most $M_0^{1-t} M_1^t$ when S is given the norm from L^{p_t}.

Proof : We let $\Omega := \{z \in \mathbf{C} \mid 0 < \Re z < 1\}$, and for $z \in \overline{\Omega}$ we put

$$\frac{1}{p(z)} = \frac{1-z}{p_0} + \frac{z}{p_1} \quad \text{and} \quad \frac{1}{q(z)} = \frac{1-z}{q_0} + \frac{z}{q_1}$$

221

Furthermore we let r_t denote the dual exponent to q_t, i.e.

$$\frac{1}{r_t} + \frac{1}{q_t} = 1$$

Clearly

$$\frac{1}{r_t} = \frac{1-t}{r_0} + \frac{t}{r_1}$$

which we extend to $z \in \overline{\Omega}$ by

$$\frac{1}{r(z)} = \frac{1-z}{r_0} + \frac{z}{r_1}$$

Note that $p(t) = p_t$, $q(t) = q_t$ and $r(t) = r_t$.

Fix $t \in\,]0,1[$. Let us for convenience write $p = p_t$, $q = q_t$ and $r = r_t$.

Let f be in S. We must show that $Tf \in L^q$ has norm less than or equal to $M_0^{1-t}M_1^t\|f\|_p$. It suffices to prove that

$$\left| \int (Tf)(x)g(x)dx \right| \le M_0^{1-t}M_1^t\|g\|_r\|f\|_p \ \ \text{for all} \ \ g \in S$$

For each $z \in \overline{\Omega}$ we define

$$f_z := sign(f)|f|^{\frac{p}{p(z)}} \ \ \text{and} \ \ g_z := sign(g)|g|^{\frac{r}{r(z)}}$$

Note that $f_t = f$ and $g_t = g$; and also that

$$|f_z| = |f_{\Re z}| = |f|^{\frac{p}{P(\Re z)}} \ \ \text{and} \ \ |g_z| = |g_{\Re z}| = |g|^{\frac{r}{r(\Re z)}}$$

If

$$f = \sum_j a_j 1_{A_j}$$

then

$$f_z = \sum_j sign(a_j)|a_j|^{\frac{p}{p(z)}} 1_{A_j}$$

so by the linearity of T we get

$$T(f_z) = \sum_j sign(a_j)|a_j|^{\frac{p}{p(z)}} T(1_{A_j})$$

Since a similar formula holds for g_z we clearly have that the function ϕ, defined on $\overline{\Omega}$ by

$$\phi(z) := \int T(f_z)g_z \ , \ \ z \in \overline{\Omega}$$

is holomorphic on Ω, and continuous and bounded on $\overline{\Omega}$. By the Three Lines Theorem applied to ϕ and the point $z = t$ we get

(1) $\ |\phi(t)| \le N_0^{1-t}N_1^t$ where $N_0 = \sup_y |\phi(iy)|$ and $N_1 = \sup_y |\phi(1 + iy)|$

222

Let us estimate $|\phi(1+iy)|$: By Hölder

$$|\phi(1+iy)| \leq \|T(f_{1+iy})\|_{q(1)}\|g_{1+iy}\|_{r(1)} \leq M_1\|f_{1+iy}\|_{p(1)}\|g_{1+iy}\|_{r(1)}$$

$$\leq M_1\|f_{\Re(1+iy)}\|_{p(1)}\|g_{\Re(1+iy)}\|_{r(1)} \leq M_1\|f\|_p^{\frac{p}{p(1)}}\|g\|_r^{\frac{r}{r(1)}}$$

so

$$N_1 \leq M_1\|f\|_p^{\frac{p}{p(1)}}\|g\|_r^{\frac{r}{r(1)}}$$

We estimate N_0 similarly. Substituting these estimates into (1) and recalling how p and r are defined we get

$$|\phi(t)| \leq M_0^{1-t}M_1^t\|f\|_p\|g\|_r$$

proving the theorem. $\qquad\qquad\qquad\qquad\qquad\qquad\qquad\qquad\qquad\qquad\qquad\qquad$ □

A bounded linear operator mapping of L^p into L^q is said to be of type (p,q). In this language Theorem 4 reads: If T is simultaneously of type (p_0, q_0) and of type (p_1, q_1), then it is also of type (p_t, q_t) for any $t \in [0,1]$ where

$$\frac{1}{p_t} = \frac{1-t}{p_0} + \frac{t}{p_1} \text{ and } \frac{1}{q_t} = \frac{1-t}{q_0} + \frac{t}{q_1}$$

and we have an estimate for the the operator norm of T.

A very simple consequence of Theorem 4 is

Theorem 5 (The Hausdorff-Young inequality).
If $p \in [1,2]$ *and* $f \in L^p(\mathbf{R}^n)$ *then its Fourier transform* Ff *belongs to* $L^{p'}(\mathbf{R}^n)$ *and*

$$\|Ff\|_{p'} \leq \|f\|_p \text{ where } \frac{1}{p'} + \frac{1}{p} = 1$$

Proof : The Fourier transformation $f \to Ff$ is simultaneously of type $(1, \infty)$ and of type $(2, 2)$. Apply the Riesz-Thorin interpolation theorem. $\qquad\qquad$ □

Section 3 M. Riesz' theorem

Any harmonic function u on the unit disc is the real part of an analytic function f (Theorem XII.2). The imaginary part v of $f = u+iv$ is not unique, but any two choices of v differ by a real constant. The *conjugate function* (or *harmonic conjugate*) of u is the v which is fixed by the requirement that $v(0) = 0$, or in other words by the requirement that $f(0)$ should be real. As simple examples we mention that the conjugate function of $u(z) := \Re(z^n)$ is $v(z) = \Im(z^n)$, and that the one of $u(z) := \Im(z^n)$ is $v(z) = -\Re(z^n)$, $n = 0, 1, \cdots$; more generally, if u is a trigonometric polynomial with real coefficients,

$$u\left(re^{i\theta}\right) = a_0 + \sum_{n=1}^{N} r^n\{a_n \cos(n\theta) + b_n \sin(n\theta)\}$$

then its conjugate function is

$$v\left(re^{i\theta}\right) = \sum_{n=1}^{N} r^n \{-b_n \cos(n\theta) + a_n \sin(n\theta)\}$$

The holomorphic functions that have u as real part are characterized by *Schwarz'* *formula*:

Theorem 6 (Schwarz' formula).

Let $f = u + iv$ be holomorphic on the open ball $B(0, R)$ and let u be continuous on the closed ball $B[0, R]$. Then

$$f(z) = iv(0) + \frac{1}{2\pi} \int_0^{2\pi} u\left(Re^{i\theta}\right) \frac{Re^{i\theta} + z}{Re^{i\theta} - z} d\theta \text{ for all } |z| < R$$

Proof : By Poisson's formula (Theorem XII.6)

$$\Re(f(z)) = u(z) = \frac{1}{2\pi} \int_0^{2\pi} u\left(Re^{i\theta}\right) \frac{R^2 - |z|^2}{\left|Re^{i\theta} - z\right|^2} d\theta$$

$$= \frac{1}{2\pi} \int_0^{2\pi} u\left(Re^{i\theta}\right) \Re\left(\frac{Re^{i\theta} + z}{Re^{i\theta} - z}\right) d\theta = \Re\left(\frac{1}{2\pi} \int_0^{2\pi} u\left(Re^{i\theta}\right) \frac{Re^{i\theta} + z}{Re^{i\theta} - z} d\theta\right)$$

so the real part of the analytic function

$$z \to f(z) - \frac{1}{2\pi} \int_0^{2\pi} u\left(Re^{i\theta}\right) \frac{Re^{i\theta} + z}{Re^{i\theta} - z} d\theta$$

is 0. Hence the function is a constant. Taking $z = 0$ we see that this constant is $iv(0)$.□

For any complex valued function ϕ on the unit disc we let

$$\|\phi\|_p := \sup_{0<r<1} \left\{ \frac{1}{2\pi} \int_0^{2\pi} \left|\phi\left(re^{i\theta}\right)\right|^p d\theta \right\}^{\frac{1}{p}} \text{ for } p \in [1, \infty[$$

With this notation a holomorphic function f on the unit disc is in H^p iff $\|f\|_p < \infty$. If f is holomorphic or harmonic then $|f|^p$ is subharmonic (Example XIII.15) and so the supremum is the limit as $r \to 1_-$ (Proposition XIII.9). If f furthermore is continuous on the closed unit disc then clearly

$$\|f\|_p^p = \frac{1}{2\pi} \int_0^{2\pi} \left|f\left(e^{i\theta}\right)\right|^p d\theta \text{ for } p \in [1, \infty[$$

i.e. the norm equals the L^p-norm of the restriction of f to the unit circle.

Let $f = u + iv$ be holomorphic on the unit disc and such that $f(0)$ is real, i.e. v is the conjugate function of u. The problem that we shall now consider is whether $\|u\|_p < \infty$ implies $\|v\|_p < \infty$, or said differently: Must $f \in H^p$ if $\|u\|_p < \infty$?

Let us analyze the special case $p = 2$, where we can compute all the norms involved. We consider an $f \in Hol(B(0,1))$,

$$f(z) = \sum_{n=0}^{\infty} a_n z^n$$

such that $f(0) = a_0 \in \mathbf{R}$. Then

$$u(z) = \Re(f(z)) = \sum_{n=0}^{\infty} \Re(a_n z^n)$$

From the theory for power series we know that the series $\sum_n a_n r^n e^{in\theta}$ for each fixed $r \in [0,1[$ converges uniformly with respect to θ to $f(re^{i\theta})$; its real part will then of course converge uniformly to $u(re^{i\theta})$. Since

$$\frac{1}{2\pi} \int_0^{2\pi} e^{in\theta} e^{-im\theta} d\theta = \delta_{n,m}$$

we get

$$\frac{1}{2\pi} \int_0^{2\pi} \left| f\left(re^{i\theta}\right) \right|^2 d\theta = \lim_{N \to \infty} \frac{1}{2\pi} \int_0^{2\pi} \left(\sum_{n=0}^{N} r^n e^{in\theta} \right) \left(\sum_{m=0}^{N} r^m \overline{a_m} e^{-im\theta} \right) d\theta$$

$$= \lim_{N \to \infty} r^{n+m} a_n \overline{a_m} \frac{1}{2\pi} \int_0^{2\pi} e^{in\theta} e^{-im\theta} d\theta = \lim_{N \to \infty} \sum_{n=0}^{N} r^{2n} |a_n|^2 = \sum_{n=0}^{\infty} r^{2n} |a_n|^2$$

so

$$\|f\|_2^2 = \sum_{n=0}^{\infty} |a_n|^2$$

In a similar way we can compute (we leave the details to the reader) that

$$\|u\|_2^2 = a_0^2 + \frac{1}{2} \sum_{n=1}^{\infty} |a_n|^2$$

In particular

$$\|f\|_2^2 = 2\|u\|_2^2 - a_0^2 \le 2\|u\|_2^2$$

so $f \in H^2$ if $\|u\|_2 < \infty$, answering our question above affirmatively.

Marcel Riesz showed in 1927 that this phenomenon persists for p in the range $]1,\infty[$:

Theorem 7 (M. Riesz' Theorem).
For each $]1,\infty[$ there exists a positive constant C_p such that

$$\|f\|_p \le C_p\|\Re f\|_p \text{ for all } f \in Hol(B(0,1)) \text{ such that } f(0) \in \mathbf{R}$$

Remark: We have not specified any bound on the constant C_p, except in the case $p = 2$. It can be shown that $C_p = (p/(p-1))^{1/p}$ works [HK].

Proof: We let for convenience $\mathcal{R}$ denote the set of those $f \in Hol(B(0,1))$ for which $f(0)$ is real.

We divide the proof into 3 steps.

Step 1: Here we prove the following claim.

Claim: Let p be fixed and assume that there is a constant $C > 0$ such that $\|f\|_p \le C\|\Re f\|_p$ for all $f \in \mathcal{R}$. Then

$$\|f\|_{2p} \le (4C+1)\|\Re f\|_{2p} \text{ for all } f \in \mathcal{R}$$

Proof of the claim: Consider first the case of $f(0) = 0$. Possibly replacing f by $z \to f(rz)$ for $r \in]0,1[$ we may assume that f is holomorphic in a neighborhood of the closed unit disc, so that

$$
\begin{aligned}
\|f\|_{2p}^{2p} &= \frac{1}{2\pi}\int_0^{2\pi}\left|f\left(e^{i\theta}\right)\right|^{2p}d\theta = \frac{1}{2\pi}\int_0^{2\pi}\left|f\left(e^{i\theta}\right)^2\right|^p d\theta \\
&= \frac{1}{2\pi}\int_0^{2\pi}\left|u^2 - v^2 + 2iuv\right|^p d\theta = \frac{2^p}{2\pi}\int_0^{2\pi}\left|uv + \frac{u^2 - v^2}{2i}\right|^p d\theta \\
&= 2^p\left\|uv + \frac{u^2 - v^2}{2i}\right\|_p^p
\end{aligned}
$$

which by our assumption this can be estimated by

$$2^p C^p\|uv\|_p^p = (2C)^p\frac{1}{2\pi}\int_0^{2\pi}|u|^p|v|^p d\theta$$

and further by the Cauchy-Schwarz inequality

$$
\begin{aligned}
&\le (2C)^p\left\{\frac{1}{2\pi}\int_0^{2\pi}|u|^{2p}d\theta\right\}^{\frac{1}{2}}\left\{\frac{1}{2\pi}\int_0^{2\pi}|v|^{2p}d\theta\right\}^{\frac{1}{2}} \\
&= (2C)^p\|u\|_{2p}^p\|v\|_{2p}^p \le (2C)^p\|u\|_{2p}^p\|f\|_{2p}^p
\end{aligned}
$$

Dividing through by $\|f\|_{2p}^p$ we get that $\|f\|_{2p} \leq 2C\|u\|_{2p}$. For a general $f \in \mathcal{R}$ we thus have established the inequality

$$\|f - f(0)\|_{2p} \leq 2C\|u - u(0)\|_{2p}$$

Now, $f(0) = u(0)$, so

$$\|f\|_{2p} \leq 2C\|u\|_{2p} + (1 + 2C)|u(0)|$$

To deal with the last term we note that $|u|^{2p}$ is subharmonic (Example XIII.15(γ)), so that

$$|u(0)|^{2p} \leq \frac{1}{2\pi} \int_0^{2\pi} \left|u\left(re^{i\theta}\right)\right|^{2p} d\theta \leq \|u\|_{2p}^{2p}$$

and finally $\|f\|_{2p} \leq (1 + 4C)\|u\|_{2p}$, proving the claim.

Step 2 :

We know that the inequality of M. Riesz' theorem holds for $p = 2$ (with the constant $1/2$). From Step 1 we see that it then also holds for $p = 4, 8, 16, \cdots$. In the present step we use the Riesz-Thorin interpolation theorem to show that the inequality holds for any p in the interval $[2, \infty[$.

The map that interests us, i.e. $u \to f = u + iv$, where v is the conjugate function of u, has according to Schwarz' formula (Theorem 5) the explicit expression

$$(*) \quad f(z) = \frac{1}{2\pi} \int_0^{2\pi} u\left(e^{i\theta}\right) \frac{e^{i\theta} + z}{e^{i\theta} - z} d\theta \quad \text{for } |z| < 1$$

if u is continuous on the closed unit disc. However, u and f are in general only defined on the open disc, so we must consider the functions on the circle $S_r := \{z \in \mathbf{C} \,|\, |z| = r\}$ and afterwards let $r \to 1_-$. More precisely, we define for each $r \in]0, 1[$ a linear map $T_r : L^1(S^1) \to C(S^1)$ by

$$(T_r\phi)\left(e^{i\theta}\right) := \frac{1}{2\pi} \int_0^{2\pi} \phi\left(e^{it}\right) \frac{e^{it} + re^{i\theta}}{e^{it} - re^{i\theta}} dt$$

and shall for each $p \in [2, \infty[$ study its restriction

$$T_{r,p} := T_r|_{L^p} : L^p(S^1) \to L^p(S^1)$$

That restriction is a continuous linear operator from $L^p(S^1)$ into $L^p(S^1)$.

Let now $\phi \in C(S^1)$ be real valued and let f denote the analytic function from (*) with u replaced by ϕ. Then

$$\|T_{r,p}\phi\|_{L^p} = \left\{ \frac{1}{2\pi} \int_0^{2\pi} \left|f\left(re^{i\theta}\right)\right|^p d\theta \right\}^{\frac{1}{p}} \leq \|f\|_p$$

If, furthermore, p is of the form $p = 2^N$ for some $N = 1, 2, \cdots$ we may by Step 1 estimate as follows:

$$\|f\|_p \leq A_p \|\Re f\|_p = A_p \|\phi\|_{L^p}$$

where A_p is a constant.

If ϕ is not real valued we split it into its real and imaginary parts to get

$$\|T_{r,p}\phi\|_{L^p} \leq 2A_p \|\phi\|_{L^p}$$

so that

$$\|T_{r,p}\|_{L^p, L^p} \leq 2A_p \quad \text{for} \quad p = 2^N, \ N = 1, 2, \cdots$$

By the Riesz-Thorin interpolation theorem there exists to each $p \in [2, \infty[$ a constant C_p, not depending on r, such that

$$\|T_{r,p}\phi\|_{L^p} \leq C_p \|\phi\|_{L^p} \quad \text{for all} \ \phi \in L^p(S^1)$$

If f is holomorphic on a neighborhood of the closed unit disc and $f(0)$ is real then

$$\|f\|_p = \lim_{r \to 1_-} \|f|_{S_r}\|_{L^p} = \lim_{r \to 1_-} \|T_{r,p}(\Re f)\|_{L^p} \leq C_p \|\Re f\|_p$$

If f is holomorphic on $B(0,1)$, but not necessarily on a neighborhood of the closed unit disc, then we apply the result just derived to the analytic function $f_\lambda(z) := f(\lambda z)$ and let $\lambda \to 1_-$.

With Step 2 we have proved M. Riesz' theorem for p in the range $[2, \infty[$.

Step 3:

Here we derive M. Riesz' theorem in the range $]1, 2[$ by the standard technique of duality, knowing it to be true in the range $[2, \infty[$.

For exponents $q < 2$ we consider the conjugate exponent $p = q/(q-1) > 2$. Let $f = u + iv \in H^q$ where $v(0) = 0$. We may assume that f is continuous on the closed disc $|z| \leq 1$ (If not consider $f(rz)$ for $r < 1$), so that the norms reduce to the relevant L^s-norms of the functions on the unit circle.

Since trigonometric polynomials are dense in $L^p(S^1)$ we have

$$\|v\|_q = \sup_g \left| \int vg \right|$$

where $\int$ denotes integration over S^1 with respect to $\frac{d\theta}{2\pi}$ and where the supremum ranges over all real trigonometric polynomials g with L^p-norm ≤ 1. Let

$$g(\theta) = a_0 + \sum_{n=1}^{N} \{a_n \cos(n\theta) + b_n \sin(n\theta)\}$$

228

be such a polynomial and consider the harmonic function given by

$$g\left(re^{i\theta}\right) = a_0 + \sum_{n=1}^{N} r^n \{a_n \cos{(n\theta)} + b_n \sin{(n\theta)}\}$$

Its conjugate function h is

$$h\left(re^{i\theta}\right) = \sum_{n=1}^{N} r^n \{-b_n \cos{(n\theta)} + a_n \sin{(n\theta)}\}$$

Since $g + ih$ is analytic so is the product $f(g + ih)$. Its imaginary part $vg + uh$ is harmonic and vanishes at 0, so by the mean value property for harmonic functions (Theorem XII.6)

$$\int vg = -\int uh$$

Combining Hölder's inequality $\left|\int uh\right| \leq \|u\|_q \|h\|_p$ with the following result, which we have already proved

$$\|h\|_p \leq \|g + ih\|_p \leq C_p \|g\|_p$$

we get

$$\left|\int vg\right| \leq C_p \|u\|_q \|g\|_p \leq C_p \|u\|_q$$

which implies that $\|v\|_q \leq C_p \|u\|_q$. Hence

$$\|f\|_q \leq \|u\|_q + \|v\|_q \leq (C_p + 1)\|u\|_q$$

which is the desired statement. $\qquad\qquad \square$

Section 4 Exercises

1. Let

$$\Omega := \left\{z \in \mathbf{C}\backslash\{0\} \mid |Arg\, z| < \frac{\pi}{2a}\right\}$$

where $a \geq \frac{1}{2}$ is a constant. Let $f \in C(\overline{\Omega}) \cap Hol(\Omega)$ satisfy that

$$|f(z)| \leq C \exp\left(A|z|^b\right) \quad \text{for all } z \in \Omega$$

where A and C are positive constants and $b \in [0, a[$.

Show that $|f| \leq M$ throughout Ω, if $|f| \leq M$ on $\partial\Omega$.

Hint: Consider the function $\phi(z) := \exp{(-z^c)}$ for $c \in \,]b, a[$.

2. Let Ω be the strip

$$\Omega := \{z \in \mathbf{C} \mid 0 < \Re z < L\}$$

where $L \in]0, \infty[$. Assume that $f \in C(\overline{\Omega}) \cap Hol(\Omega)$ satisfies that

$$|f(z)| \leq C \exp\left(Ae^{a|z|}\right) \quad \text{for} \ z \in \Omega$$

where A and C are positive constants and $a \in [0, \frac{\pi}{2L}[$.

Show that $|f| \leq M$ throughout Ω, if $|f| \leq M$ on $\partial\Omega$. Hint: Consider for $b \in]a, \frac{\pi}{2L}[$ the function

$$\phi(z) := \exp\left(-e^{-ibz} - e^{ibz}\right)$$

References

Aa B. Åkeberg, Proof of Poisson's Formula. Proc. Camb. Phil. Soc. **57** (1961), 186.

AG B. Artmann and W. Gerecke, Didaktische Beobachtungen zu einem Divergenzbeweis für $\Sigma 1/p$ von Paul Erdös. Math. Semesterber. **37** (1990), 46–56.

AGR A.K. Arora, S.K. Goel and D.M. Rodriguez, Special Integration Techniques for Trigonometric Integrals. Amer. Math. Monthly **95** (1988), 126-130.

An Anonyme, Sur l'intégrale $\int_0^\infty e^{-x^2}dx$. Bull. Sci. Math. II **13** (1889), 84.

Ap1 T.M. Apostol : "Introduction to Analytic Number Theory" Springer-Verlag 1976.

Ap2 T.M. Apostol, A Proof that Euler Missed: Evaluating $\zeta(2)$ the Easy Way. Math. Intelligencer **5** no 3 (1983), 59–60.

Ax S. Axler, Harmonic Functions from a Complex Analysis Viewpoint. Amer. Math. Monthly **93** (1986), 246-259.

Ay R. Ayub, Euler and the ζ-function. Amer. Math. Monthly **81** (1874), 1067–1086.

Be A. Beck, The Broken Spiral Theorem. Amer. Math. Monthly **93** (1986), p. 293.

Ber B.C. Berndt, Elementary evaluation of $\zeta(2n)$. Math. Magazine **48** (1975), 148–154.

Bo H.P. Boas and R.P. Boas, Short Proofs of Three Theorems on Harmonic Functions. Proc. Amer. Math. Soc. **102** (1988), 906-908.

Bu R.B. Burckell : "An Introduction to Classical Complex Analysis" Birkhäuser 1979.

Ca D.M. Campbell, Beauty and the Beast: The Strange Case of André Bloch. Math. Intelligencer **7** no 4 (1985), 36-38.

References

Ch P.R. Chernoff, Liouville's Theorem for Harmonic Functions. Amer. Math. Monthly **79** (1972), 310–311.

CF H. Cartan and J. Ferrand, The Case of André Bloch. Math. Intelligencer **10** no 1 (1988), 23-26.

CS J. Crowe and D. Samperi, On Open Maps. Amer. Math. Monthly **96** (1989), 242-243.

Dn K.R. Davidson, Pointwise limits of analytic functions. Amer. Math. Monthly **90** (1983), 391–394.

Ds P.C. Davis, Leonhard Euler's Integral: A Historical Profile of the Γ-Function. Amer. Math. Monthly **66** (1959), 849-869.

DG K.K. Dewan and N.K. Gowil, On the Eneström-Kakeya Theorem. Jour. Approx. Theory **42** (1984), 239-244.

Di J.D. Dixon, A brief proof of Cauchy's integral theorem. Proc. Amer. Math. Soc. **29** (1971), 625–626.

DS N. Dunford and J.T. Schwartz : "Linear Operators. Part I : General Theory" Interscience Publishers, Inc. New York. 1958.

Ed H.M. Edwards : "Riemann's Zeta Function" Academic Press. New York and London 1974.

Ga P.M. Gauthier, Remarks on a theorem of Keldysh-Lavrent'ev. Russ. Math. Surv. **40** (1985), 179-180.

Gl J. Globevnik, Zero integrals on circles and characterizations of harmonic and analytic functions. Trans. Amer. Math. Soc. **317** (1990), 213-330.

GM J.D. Gray and S.A. Morris, When is a function that satisfies the Cauchy-Riemann equations analytic? Amer. Math. Monthly **85** (1978), 246–256.

Ha P.R. Halmos, Does Mathematics Have Elements ? Math. Intelligencer **3** (1980), 147-153.

HK W.K. Hayman and P.B. Kennedy : "Subharmonic Functions" Vol I. Academic Press. London. New York. San Francisco. 1976.

References

HI E. Hille : "Analytic Function Theory" I. Blaisdell 1959.

HII E. Hille : "Analytic Function Theory" II. Ginn and Company 1962.

Hö L. Hörmander : "An introduction to COMPLEX ANALYSIS IN SEVERAL VARIABLES" North Holland 1973.

Kn K.Knopp : "Theory and Application of Infinite Series" Blackie & Son Ltd, Glasgow (2nd ed. 1951).

Le A. Lenard, A Note on Liouville's Theorem. Amer. Math. Monthly **93** (1986), 200-201.

LR J. van de Lune and H.J.J. te Riele, On the zeros of the Riemann zeta function in the critical strip.III. Math. Comp. **41** (1983),759–767.

Mac T.H. MacGregor, Geometric Problems in Complex Analysis. Amer. Math. Monthly **79** (1972), 447-468.

Me M.D. Meyerson, Every Power Series is a Taylor Series. Amer. Math. Monthly **88** (1981), 51-52.

Mi D. Minda, The Dirichlet problem for a disk. Amer. Math. Monthly **97** (1990), 220-223.

MS D. Minda and G. Schober, Another Elementary Approach to the Theorems of Landau, Montel, Picard and Schottky. Complex Variables **2** (1983), 157-164.

Na V.S. Narasimhan : "Complex Analysis in One Variable" Birkhäuser 1985.

Neu E. Neuenschwander, The Casorati-Weierstrass theorem (studies in the history of complex function theory I). Historia Math. **5** (1978), 139-166. Zbl. 374 # 01010.

New D.J. Newman, Fourier Uniqueness via Complex Variables. Amer. Math. Monthly **81** (1974), 379-380.

No W.P. Novinger, Some Theorems from Geometric Function Theory : Applications. Amer. Math. Monthly **82** (1975), 507-510.

OS A.V. Oppenheim and R.W. Schafer : "Digital Signal Processing" Prentice-Hall, Inc., Englewood Cliffs, New Jersey 1975.

References

Pa J.M. Patin, A Very Short Proof of Stirling's Formula. Amer. Math. Monthly **96** (1989), 41-42.

Pe I.G. Petrowski : "Vorlesungen über partielle Differentialgleichungen" B.G. Teubner Verlagsgesellschaft. Leipzig 1955.

Po A.v.d. Poorten, A Proof that Euler Missed ... Apéry's Proof of the Irrationality of $\zeta(3)$. Math. Intelligencer **1** (1979), 195–203.

PW M.H. Protter and H.F. Weinberger : "Maximum Principles in Differential Equations" Prentice-Hall, Inc. Englewood Cliffs, N.J. 1967.

Ra K.N. Srinivasa Rao, A Contour for the Poisson Integral. Elem. Math. **27** (1972), 88-90.

Ro J.-P. Rosay, Injective Holomorphic Mappings. Amer. Math. Monthly **89** (1982), 587-588.

RR J.-P. Rosay and W. Rudin, Arakelian's Approximation Theorem. Amer. Math. Monthly **96** (1989), 432-433.

Ru W. Rudin : "Real and Complex Analysis" 2nd ed. TMH-edition 1974, Tata McGraw-Hill 1974.

Se A. Selberg, An elementary proof of the prime number theorem. Ann. of Math. **50** (1949), 305-313.

SG G.Sansone and J.Gerretsen : "Lectures on the theory of functions of a complex variable" I-II. P. Noordhoff-Groningen 1960.

Sp E.H. Spanier : "Algebraic Topology" McGraw-Hill 1966.

Ti M. Tideman, Elementary proof of a uniqueness theorem for positive harmonic functions. Nordisk Mat. Tidsskrift **2** (1954), 95–96.

Tr F. Treves : "Basic Linear Partial Differential Equations" Academic Press 1975.

Ts A. Tsarpalias, A Version of Rouché's Theorem for Continuous Functions. Amer. Math. Monthly **96** (1989), 911–913.

References

Vi A.G. Vitushkin, Uniform approximation of functions by holomorphic functions. Proc. Steklov Inst. Math. **176** (1988), 301–308.

Wag S. Wagon, Where are the zeros of zeta of s? Math. Intelligencer **8** no 4 (1986), 57–62.

Wal W. Walter, Old and new approaches to Euler's trigonometric expansions. Amer. Math. Monthly **89** (1982), 225–230.

Yo R.M. Young, On Jensen's Formula and $\int_0^{2\pi} \log|1 - e^{i\theta}|d\theta$. Amer. Math. Monthly **93** (1986), 44-45.

Za L. Zalcman, Real Proofs of Complex Theorems (and Vice Versa). Amer. Math. Monthly **81** (1974), 115-137.

Index

A

Abel's partial summation formula 7
Abel's test 11
Abel's theorem 5, 11
addition theorem 23
Ahlfors' lemma 110, 112
alternating series theorem 11
analytic 16
angle 118
annulus 71, 80
argument 25
argument principle 129, 144

B

Bernoulli's paradox 38
Blaschke product 166
Bloch-Landau's theorem 101
Bloch's constant 101
Bloch's theorem 101
Borsuk-Ulam's theorem 42
branch 25, 35
Brouncker's series 6
Brouwer's fixed point theorem 30, 67
Brouwer's theorem on invariance
  of domain 41

C

Carleman's theorem 139
Casorati-Weierstrass' theorem 100
Cauchy inequality 11
Cauchy integral formula 47, 71
Cauchy integral formula for a disc 48
Cauchy integral theorem 47, 71, 72, 74
Cauchy-Goursat 43
Cauchy-Green formula 48

Cauchy-Hadamard radius of
  convergence formula 2
Cauchy-Riemann equation 14
Cayley transform 119
Cayley transformation 122
chain rule 13
Chebyshev's function 175
circle 114, 117
circle of convergence 2
closed curve 17
complementary arguments 158
complex derivative 13
complex differentiable 13
condensation principle 120
conformally equivalent 122
conjugate function 223
continuous argument 25
continuous logarithm 25
contractible 38
converge 147
converge locally uniformly 149
critical strip 176
curvature 109
curve 17

D

derivative 13
Dirichlet problem 190
duplication formula of
Legendre 161, 167

E

Eneström-Kakeya 10
entire 13
essential singularity 83, 92, 115
Euler's constant 157, 167